COMMUNICATING CLIMATE CHANGE

A VOLUME IN THE SERIES

Cornell Studies in Environmental Education

Edited by Marianne E. Krasny

For a list of books in the series, visit our website at cornellpress.cornell.edu.

COMMUNICATING CLIMATE CHANGE

A Guide for Educators

Anne K. Armstrong,
Marianne E. Krasny,
and Jonathon P. Schuldt

COMSTOCK PUBLISHING ASSOCIATES
AN IMPRINT OF
CORNELL UNIVERSITY PRESS   ITHACA AND LONDON

First published 2018 by Cornell University Press

Printed in the United States of America

Library of Congress Cataloging-in-Publication Data

Names: Armstrong, Anne K., author. | Krasny, Marianne E., author. | Schuldt, Jonathon P., author.
Title: Communicating climate change : a guide for educators / Anne K. Armstrong, Marianne E. Krasny, and Jonathon P. Schuldt.
Description: Ithaca [New York] : Cornell University Press, 2018. | Series: Cornell studies in environmental education | Includes bibliographical references and index.
Identifiers: LCCN 2018030922 (print) | LCCN 2018036694 (ebook) | ISBN 9781501730801 (pdf) | ISBN 9781501730818 (ret) | ISBN 9781501730795 | ISBN 9781501730795 (pbk. ; alk. paper)
Subjects: LCSH: Communication in climatology—United States. | Climatic changes—Study and teaching—United States. | Environmental education—United States.
Classification: LCC QC902.9 (ebook) | LCC QC902.9 .A76 2018 (print) | DDC 363.738/74071073—dc23
LC record available at https://lccn.loc.gov/2018030922

Dedicated to my family
—A.K.A.

Contents

Acknowledgments

We would like to thank Glen Koehler and Michael Hoffmann for their thorough review and helpful suggestions on climate science. We would also like to thank Adam Ratner, Jennifer Hubbard-Sanchez, Maria Talero, Karen Temple-Beamish, and Laura Mack for their time spent talking with author Anne Armstrong and for their dedication to developing innovative climate change education programs. This publication was funded in part by the U.S. Environmental Protection Agency (EPA, Assistant Agreement No. NT-83497401) and U.S. Department of Agriculture (USDA) National Institute for Food and Agriculture funds awarded to Cornell University (Award No. 2016-17-215). Neither EPA nor USDA has reviewed this publication. The views expressed are solely those of the authors. Finally, the authors gratefully acknowledge the support of Cornell University Library in enabling publication of this volume on an open access basis.

COMMUNICATING CLIMATE CHANGE

INTRODUCTION

"If only they knew more about the issue, they would act!" Have you said that to yourself or your environmental education colleagues before? Looking at an issue like climate change, we see that a wealth of information and a high level of issue awareness among the U.S. public have not led to the kind of action needed to reduce climate threats to human and natural systems. Americans' climate change concern still ranks lower than their concern for other environmental problems like water supply and pollution, as well as lower than their concern for health care and the economy. Climate change concern has, however, increased significantly since 2015.[1] Yet these high levels of awareness and growing concern mask the range of opinions that environmental educators might encounter at a local level, as well as the emergence of climate change as a highly politicized issue in U.S. politics.[2] Although climate change remains a challenging topic for environmental educators, environmental education is an important player in fostering positive climate change dialogue and subsequent climate change action.[3]

Environmental education programs, organizations, and online resources related to climate change abound in formal, nonformal, and informal settings.[4] The Climate Literacy and Energy Awareness Network (CLEAN) boasts a collection of over six hundred climate change education resources reviewed by scientists and educators that range from activities to demonstrations, visualizations, and videos curated from around the Internet. National environmental education training programs like Project Learning Tree focus their attention on climate change, with a module for secondary education called *Southeastern Forests and Climate Change*.[5] The National Network for Ocean and Climate Change

Interpretation (NNOCCI) has trained over 150 educators in thirty-eight states in research-based techniques for engaging audiences with climate change. And the Planet Stewards program of the National Oceanographic and Atmospheric Administration (NOAA) offers face-to-face training for educators, as well as a webinar series on climate change science and education. As interest from environmental educators has grown, so has research on developing effective climate change programs, particularly in formal education settings.[6]

Yet the question remains: How do we optimize programs for attaining climate literacy and action to address mitigation of greenhouse gas emissions, and, when necessary, adaptation to changes already taking place? A review of climate change education literature focused on education in formal settings found that making climate change "personally relevant and meaningful," and engaging learners through inquiry and constructivist learning, correlated with a program's success in increasing climate science understanding, shifting climate change attitudes, and inspiring action.[7] Research from environmental psychology and climate change communication offers useful, tangible insights into designing climate change education programs that are personally relevant and meaningful.[8] For example, environmental psychology informs climate change communication research on framing and metaphors, and it can also directly inform how educators think about and assess their audiences (figure i.1). Similarly, climate change communication research on framing can inform environmental educators' strategic choice of program language. Training programs like NNOCCI have adopted evidence-based methods drawn from climate change communication and environmental psychology, and educators who participate in this program adopt research-based practices and value a research-based approach.[9]

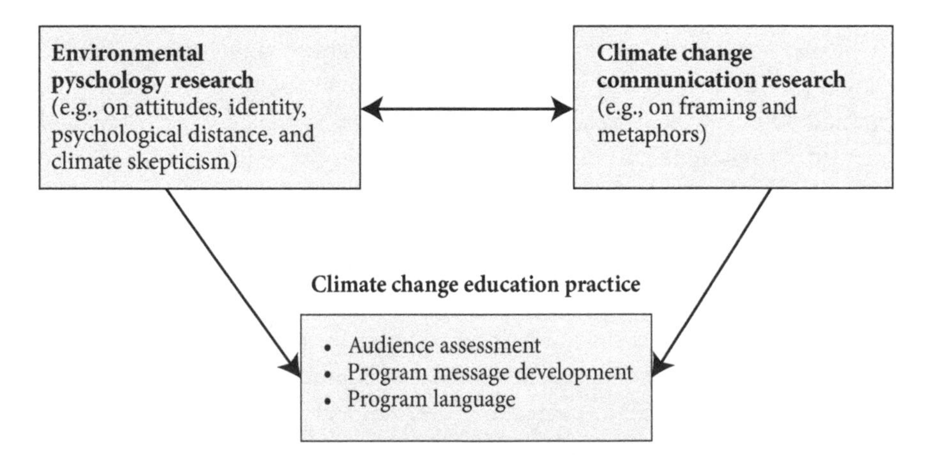

FIGURE i.1 How environmental psychology research and climate change communication research can inform climate change education practice

Climate change education and climate change communication share similar goals and desired outcomes, and their definitions reflect these similarities. Climate change education, or climate change environmental education, encompasses a range of "interdisciplinary learning opportunities that people of all ages need to develop the competencies, dispositions and knowledge to address climate change." It approaches climate change with an "understanding of the socio-political and economic considerations; the scientific basis; and the communication, collaborative problem-solving and analytical skills needed to generate and implement feasible solutions."[10] According to the Yale Program on Climate Change Communication, climate change communication is "about educating, informing, warning, persuading, mobilizing and solving this critical problem. At a deeper level, climate change communication is shaped by our different experiences, mental and cultural models, and underlying values and worldviews."[11] The first part of this definition speaks to goals held in common between climate change communication and environmental education, like climate literacy and action, while the second part touches on linkages between climate change communication and environmental psychology.

This book seeks to provide environmental educators with an understanding of how their audiences engage with climate change information, as well as with concrete, empirically tested communication tools they can use to enhance their climate change programs. We define "environmental educator" broadly, to mean people "focused on using best practice in education . . . to address the social and environmental issues facing society."[12] We focus primarily on the first three steps of developing a climate change education program (figure i.2): identifying climate change education outcomes and resources, assessing audiences, and strategizing programs. Part 1 of this book provides overviews of climate change science, climate change attitudes and knowledge, and climate change education outcomes. It also introduces three vignettes referenced throughout the chapters describing how fictional educators address climate change education challenges. Part 2 explores how psychology research explains the complex ways in which people interact with climate change information; this research is useful in informing educators' audience assessment. Part 3 presents communication strategies with a focus on research about framing, metaphors, and messengers that can help educators formulate program language. At the end of parts 2 and 3, we summarize the research with an eye toward applications to environmental education. Finally, part 4, "Stories from the Field," highlights four educators' climate change education programs and illustrates connections between their teaching strategies and the research covered in parts 2 and 3.

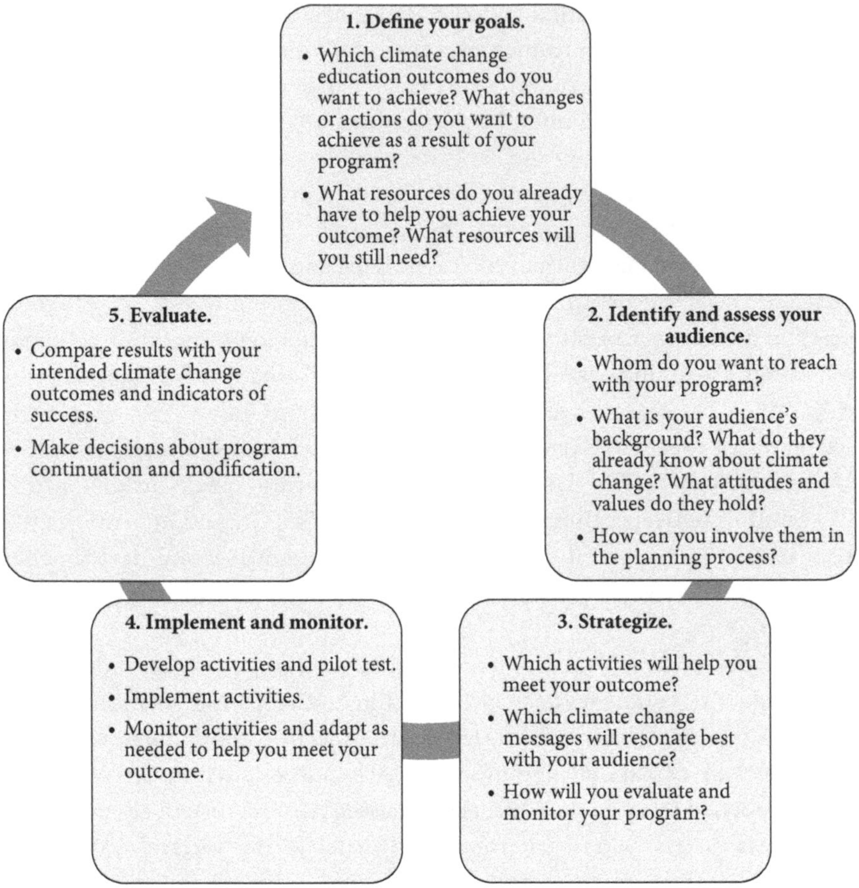

1. Define your goals.

- Which climate change education outcomes do you want to achieve? What changes or actions do you want to achieve as a result of your program?
- What resources do you already have to help you achieve your outcome? What resources will you still need?

2. Identify and assess your audience.

- Whom do you want to reach with your program?
- What is your audience's background? What do they already know about climate change? What attitudes and values do they hold?
- How can you involve them in the planning process?

3. Strategize.

- Which activities will help you meet your outcome?
- Which climate change messages will resonate best with your audience?
- How will you evaluate and monitor your program?

4. Implement and monitor.

- Develop activities and pilot test.
- Implement activities.
- Monitor activities and adapt as needed to help you meet your outcome.

5. Evaluate.

- Compare results with your intended climate change outcomes and indicators of success.
- Make decisions about program continuation and modification.

FIGURE i.2 Program development cycle

Adapted from Susan Jacobson, *Communication Skills for Conservation Professionals*, 2nd ed. (Washington: Island Press, 2009), 50–51

Bottom Line for Educators

The complexity of climate science combined with the complicated political and cultural contexts in which people live makes climate change a particularly challenging topic to approach no matter the educational setting. This book introduces environmental psychology and climate change communication research that can assist environmental educators at several program development stages. Of course, educators also need a foundation in climate change science, which is where we turn next.

Part 1
BACKGROUND

In part 1, we begin with a chapter on how climate change works and how we know the climate is changing. Chapter 1 also includes examples of climate change actions directed at the largest sources of greenhouse gases. Chapter 2 summarizes research on climate change attitudes and knowledge. Chapter 3 outlines a variety of climate change education outcomes to assist educators in defining what they want to achieve with their programs. Chapter 4 presents three vignettes of fictional climate change educators, Elena, Jayla, and Will, who conduct programs in different settings with different audiences. Together, these four chapters provide background and context for the environmental psychology and communications research presented in parts 2 and 3.

1

CLIMATE CHANGE SCIENCE

The Facts

In this chapter, we present a short summary of weather and climate as well as an overview of climate change causes, evidence, and impacts. We also introduce actions needed to reduce greenhouse gas emissions, thus mitigating climate change. Because environmental educators know their communities, they can play a key role in distilling scientific information and guiding discussion about complexities associated with weather, climate, and climate change. They can also lead their students and communities in taking meaningful action to reduce greenhouse gases.

Weather and Climate

Weather varies minute to minute, hour to hour, day to day, month to month, and season to season. Temperatures go up and down; some days are cloudy and rainy, while others are sunny; and sometimes the air is still, whereas other times we are refreshed by a gentle breeze or buffeted about by a strong wind. Occasionally, we get floods or droughts.

In contrast to the short-term atmospheric changes we call weather, climate refers to longer-term variations. We can think of climate as the *average* weather for a particular region and time period, usually over thirty years. For example, increases in average temperatures over decades provide evidence of a changing climate. Looking to the future, scientific climate models predict longer and more severe periods of dry weather in some regions, while other regions will likely experience an increase in annual precipitation, as well as more severe rain events. In 2017, warmer and wetter atmospheric conditions and warmer ocean temperatures intensified Hurricanes Harvey, Irma, and Maria in the eastern United States, while dry weather exacerbated California wildfires—all the result of a warming planet. The more extreme weather events that we are experiencing currently will likely only intensify as average global temperatures continue to rise.

Greenhouse Gases and Climate Change

Humans, like all life on earth, depend on energy coming from the sun. But we also depend on the energy reflected from the earth's surface back into the atmosphere. This balance between energy coming in and energy going out has been maintained for billions of years, allowing life on earth to survive and thrive.

But what happens if excess greenhouse gases in the earth's atmosphere block more energy from leaving the atmosphere, upsetting that balance? What if, instead of leaving the atmosphere and going back into space, some of the excess energy is returned to the earth's surface? Put simply, the surface of the earth—including its oceans, land, and air—heats up.

Greenhouse gases are essential to life on earth. For example, plants depend on carbon dioxide (CO_2), which is also an important greenhouse gas contributing to global warming. And greenhouse gases help to maintain the earth's surface and oceans at temperatures that enable life to flourish on our planet. But as greenhouse gases accumulate beyond their historic levels, they prevent more and more of the energy reaching the earth from going back into space.

The earth absorbs sunlight energy and reemits it as heat, or what scientists call long-wave infrared radiation. Imagine this infrared radiation heading toward space. It bumps into gases in our atmosphere, like oxygen and nitrogen, and continues on its way. But if it bumps into a molecule of a greenhouse gas—say CO_2—that molecule absorbs the infrared radiation coming from the earth's surface. The molecule of CO_2 then vibrates and releases heat. The heat from the molecule can go in any direction, including up toward space or back down toward the earth.

So far, no problem. Some heat radiates out to space, and some warms up the atmosphere, oceans, and land surface (figure 1.1). But when humans start changing the balance of gases in the atmosphere—specifically, by significantly increasing the concentration of CO_2 and other greenhouse gases—more heat is emitted, including heat headed back toward the earth's surface. This leads to warming of the atmosphere, the oceans, and the land surfaces.

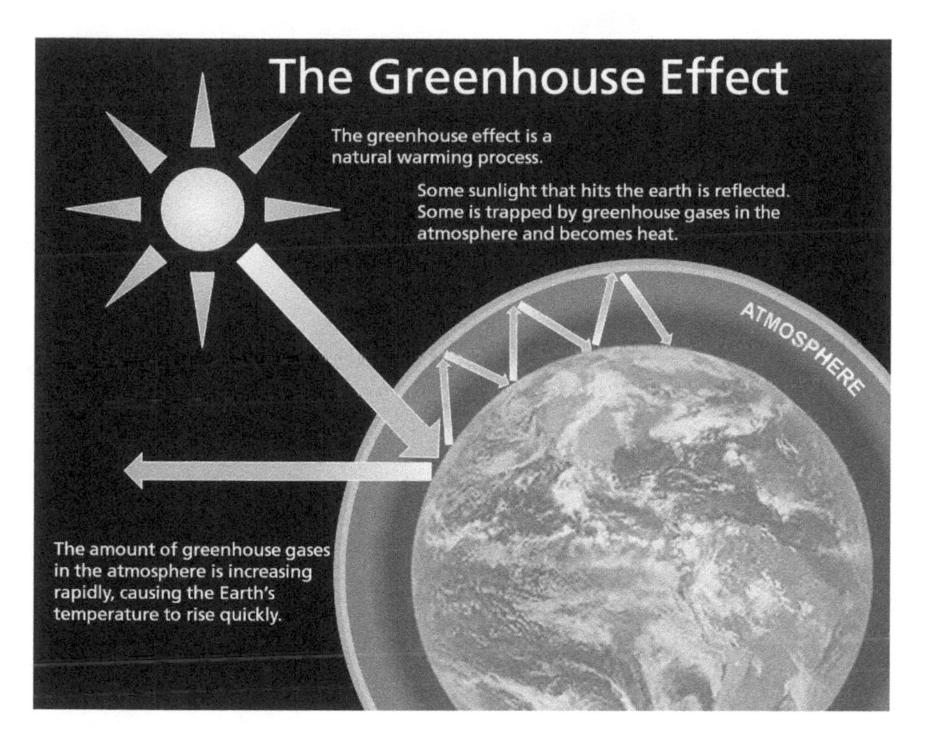

FIGURE 1.1 The greenhouse gas effect
Lindsay Modugno, Jeff Pace, and Dan Lidor, "The Effects of Climate Change and Sea Level Rise on the Coast," Sandy Hook Cooperative Research Programs, January 2015

To help people envision this process, scientists have used the analogy of a blanket surrounding the earth. On a cold night, you sleep under a blanket, and your body generates heat. The blanket traps that heat, allowing you to sleep through the night. But if your blanket is too thick, it may trap too much heat, and you start sweating and feel uncomfortable. So you can imagine the earth as being wrapped in a blanket of greenhouse gases that is trapping more heat.

So what are these greenhouse gases, and where do they come from? The most common greenhouse gas is carbon dioxide, or CO_2, which accounted for 82 percent of U.S. greenhouse gas emissions by weight in 2015 (figure 1.2). When we burn fossil fuels like coal, natural gas, and oil, which consist largely of carbon, the carbon combines with oxygen to form CO_2. Other sources of CO_2 include burning wood and decomposition of solid waste. Cement manufacturing is another significant source of greenhouse gases, accounting for 5 percent of global CO_2 emissions.[1]

Other greenhouse gases are less common but more potent than CO_2—that is, they absorb and release more heat per pound emitted. Methane accounted

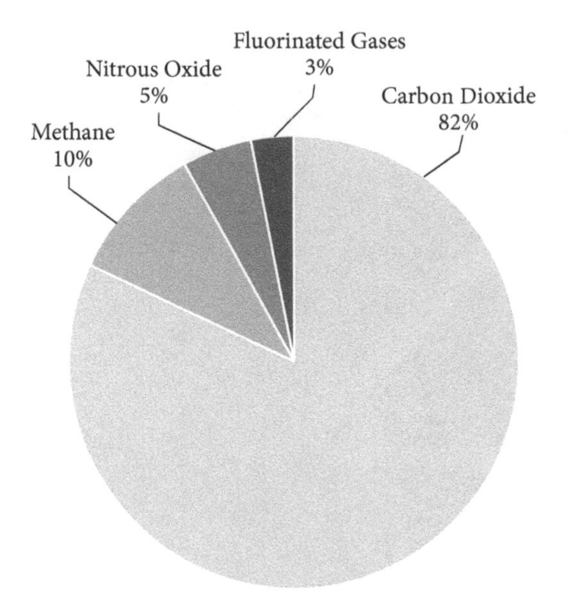

FIGURE 1.2 U.S. greenhouse gas emissions in 2015
U.S. Environmental Protection Agency, 2017

for 10 percent of U.S. greenhouse gas emissions in 2015. Methane (CH_4) is emitted in the mining and transport of natural gas, by livestock, through rice cultivation and other farming practices, and when organic waste in landfills decomposes. Similarly, nitrous oxide (N_2O), 5 percent of emissions, is emitted by agricultural and industrial activities, burning fossil fuels, and solid waste decomposition. Finally, fluorinated gases are produced by some industries and have the highest global warming potentials. Whereas methane is about thirty times more potent as a greenhouse gas relative to CO_2, nitrous oxide is nearly three hundred times as potent, and fluorinated gases can be thousands or even tens of thousands of times more potent.[2]

In fact, scientists have known about the heating effect of CO_2 since the 1850s, when the scientist John Tyndall conducted meticulous experiments on the ability of atmospheric gases to absorb and transmit radiant heat.[3] He found that CO_2 absorbed heat more readily than other atmospheric gases, like oxygen and nitrogen, which have simpler molecular structures relative to CO_2. Tyndall also speculated that small changes in gasses that absorbed the sun's heat "would produce great effects on the terrestrial rays and produce corresponding changes of climate"[4]—something that has since come to pass.

But even before Tyndall, Eunice Foote conducted an experiment in which she placed cylinders containing CO_2 and normal air in the sun and compared their temperatures. Just as Tyndall grasped the connection between CO_2 heating up faster than other gases, Foote wrote about CO_2: "An atmosphere of that gas would give to our earth a high temperature; and if as some suppose, at one period of its history the air had mixed with it a larger proportion than at present, an increased temperature from its own action as well as from increased weight must have necessarily resulted."[5]

It appears that Foote was not allowed to present her work at a scientific conference, as female presenters were uncommon in that era. Instead, in 1856, Professor Joseph Henry presented Foote's work at the meetings of the American Association for the Advancement of Science in Albany, New York, where he prefaced his explanation by pointing out that science is "of no country and of no sex."[6] More recently, researchers discovered that Foote herself published a short paper outlining her results recounting how the CO_2 container (known at the time as "carbonic acid gas")

> became itself much heated—very sensibly more so than the other—and on being removed, it was many times as long in cooling. . . .
>
> . . . On comparing the sun's heat in different gases, I found it to be in hydrogen gas, 104°; in common air, 106°; in oxygen gas, 108°; and in carbonic acid gas, 125°.[7]

In short, thanks to the experiments of Foote and Tyndall, we have known for over a century and a half about the connection between CO_2 and heating of the atmosphere.

Evidence of Climate Change

So far, we have explored the mechanisms for how greenhouse gases trap heat. But what is the evidence that the earth's climate is heating up? And even if it is warming, how do we know that factors other than greenhouse gases are not responsible? The evidence comes from measurements of greenhouse gases in the atmosphere and of recent and historical changes in the earth's surface temperature.

Between 1970 and 2000, total greenhouse gas emissions from human activities like burning fossil fuels increased an average of 1.3 percent each year. Between 2000 and 2010, total emissions increased an average of 2.2 percent per year. While this may not seem like a lot, it is similar to compound interest rates—a little bit each year can mean big changes over multiple years.

In the year 1970, humans emitted twenty-seven billion tons of greenhouse gases into the atmosphere, whereas by 2010, we emitted forty-nine billion tons of greenhouse gases per year.[8] Focusing just on CO_2, in 1850, around the time

Foote and Tyndall were conducting their experiments, the average CO_2 concentrations in the atmosphere were about 280 ppm (parts per million).[9] As of 2016, the global average CO_2 level in atmosphere was 403 ppm and increasing by 2–3 ppm per year. The last time earth's atmospheric CO_2 concentration exceeded 400 ppm was three to five million years ago, a time when global temperatures were 2° to 3°C warmer and sea levels were ten to twenty meters higher than today.[10]

Just since the late nineteenth century, the planet's average surface temperature has risen about 1.1°C (2.0°F). The current rate of warming is roughly ten times faster than the average rate of warming after ice ages of the past million years.[11] And for each decade since 1950, the global average land and ocean surface temperatures have been warmer than those for the preceding decade.[12] Temperatures are increasing faster over land and in the Northern Hemisphere than over the ocean and in the Southern Hemisphere. Temperatures are increasing fastest in the high northern latitudes such as Alaska, northern Canada, northern Russia, and across the Arctic.

Could these changes be the result of natural shifts in the earth's climate? A number of natural processes cause the earth's climate to change over time. Variations in the earth's tilt and orbit around the sun, called Milankovitch cycles, change the earth's climate over the course of tens or hundreds of thousands of years by impacting how much solar radiation reaches the earth.[13] Additionally, the El Niño and La Niña ocean warming and cooling cycle impacts temperatures and rainfall in places around the world.[14] These patterns still affect earth's climate today, but their influence over decades or even centuries is very small, much smaller than the rate of change we are now measuring. In short, these natural patterns do not explain the rapid warming that the earth has experienced since the onset of the Industrial Revolution.[15] Instead, we know from multiple sources of evidence—including long-term observations, experiments, modeling, and measurements showing that recent changes in weather patterns fit with the predictions of greenhouse gas climate change models—that increases in human-emitted greenhouse gases are responsible for climate change.

Interestingly, some natural processes also result in cooling of the earth's climate. In 1783, while he was serving as a diplomat in Paris, Benjamin Franklin observed that both Europe and the United States experienced unusually cold temperatures, as well as a constant fog. Although Franklin may not have discerned the cause, we now know that catastrophic volcanic eruptions in Iceland not only rained acid on the island itself, devastating livestock and causing widespread famine, but also caused cooling in Europe and North America. Volcanic eruptions spew tiny ash particles into the atmosphere, which decrease the amount of sunlight reaching the surface of the earth, thus lowering average global temperatures. Volcanoes that release large quantities of sulfur dioxide have an even greater effect on global temperatures; the sulfur dioxide combines with water to make a haze of tiny droplets of sulfuric acid that absorb incoming solar radiation and scatter it back out into space, thus cooling the earth's surface. Scientists today are reconstructing the history of earth's climate using tree rings and other data sources and have noted multiple periods of cooler temperatures following volcanic eruptions, which they refer to as "little ice ages."[16] However, scientists do not expect such volcanic eruptions to counteract the effects of greenhouse gas emissions.

Climate Change Impacts

In addition to scientists, many people whose lives and livelihoods are affected by changes in our oceans and on land have observed the impacts of climate change. These include coastal residents, farmers, fishermen, and leaders in the armed services. In this section, we briefly review some of these impacts.

Ocean Waters Are Becoming More Acidic

About one-quarter of the CO_2 humans produce each year is absorbed by oceans. This CO_2 reacts with seawater to form carbonic acid, thereby increasing the ocean's acidity. Similar to how the rate of CO_2 accumulation in the atmosphere is many times faster than we have seen during other periods in earth's history, the current rate of increase in the acidity of ocean surface waters is roughly fifty times faster than known historical change.[17]

What happens to sea life as the oceans acidify? The increase in carbonic acid makes calcium carbonate less available to marine organisms for building their shells. Corals, crabs, clams, oysters, lobsters, and other marine animals that form calcium carbonate shells are particularly vulnerable. Because these animals are often at the bottom of the food web, this impacts other animals, including humans.

Ocean Temperatures Are Rising

In addition to absorbing CO_2, oceans absorb heat caused by emissions from human activity. Over 90 percent of earth's warming over the past fifty years has occurred in the oceans, which have warmed 1.0°C (1.5°F) since the late nineteenth century. Rising ocean temperatures are disrupting fish populations and killing off coral reefs, in turn impacting ocean food webs, humanity's food supply, jobs, and tourism.[18]

Ice Is Melting

Glaciers in countries around the world and sea ice at the poles are melting. On average, Arctic sea ice now starts melting eleven days earlier and refreezing twenty-six days later than it did in the late 1970s. In October 2017, the volume of Arctic sea ice was 65 percent below the maximum October ice volume in 1979. Although Antarctica had been gaining ice from the 1970s to 2016, this gain was more than offset by annual losses of Arctic sea ice. Then, in 2017, Antarctic sea ice decreased to record lows.[19]

Ice loss impacts Arctic peoples who depend on traditional weather patterns for hunting and threatens animals that inhabit the Arctic. But most people don't live near glaciers and sea ice (one reason why using an image of a polar bear to inspire climate action has not been particularly effective). So why should people who do not live in icy places on the planet care about loss of glaciers and sea ice? Both melting glaciers and polar land ice cause sea level rise. Further, the loss of glaciers in the Himalayas and other mountain ranges results in changes in water flow into rivers such as the Ganges, which millions of people depend on for their water supply.[20]

Sea Level Is Rising

As glacial and polar ice melts from land, more water flows into the oceans. As water warms, it expands in volume. Both more water and warmer water are causing sea level rise. Between 1880 and 2014, sea level rose about 8 inches; by 2100, scientists are predicting an increase of 1–4 feet (0.3–1.2 meters) over the 2014 global average level, with potential for a rise of 8 feet (2.4 meters) or more if greenhouse gas emissions continue increasing. This sea level rise is not distributed evenly around the world. For example, because of ocean currents, land subsidence, and other factors, the rate of sea level rise for the East Coast of the United States is about 50 percent higher than the global average.[21]

A July 2016 headline in the *Navy Times* reads: "Rising oceans threaten to submerge 128 military bases." Norfolk, Virginia, home to the largest U.S. naval base, is already witnessing regular flooding, forcing residents to abandon their homes.[22] Frequent coastal flooding is making it nearly impossible for Norfolk residents to insure—let alone sell—once-valuable oceanfront homes. And at the Naval Academy in Annapolis, Maryland, classrooms, dormitories, and athletic facilities were flooded in a 2003 hurricane, pointing not only to sea level rise but also to stronger storm events as threats to coastal cities.[23] Residents in coastal Alaska and Louisiana, and on islands from the Pacific Ocean to the Chesapeake Bay, are abandoning villages and even whole islands where they have lived for centuries.[24]

Storm surges can cause widespread coastal property damage and kill people during a hurricane. A storm surge is the rise in ocean water above the normal tide due to a storm and is a major cause of flooding in hurricanes. It is caused by water being pushed toward the shore by storm winds. Although many factors, including water depth near the shoreline, impact storm surges, larger storms produce higher surges.[25]

Local and Regional Weather Is Changing

Recent droughts in the western United States are the most severe in over eight hundred years. At the same time, heavy rains associated with warming trends are contributing to more frequent and larger floods. Summer temperatures have exceeded those recorded since the United States began keeping reliable records in the late 1800s. And the length of the growing season between the latest spring frost and earliest fall frost has increased in each region of the United States, with increases of six days in the Southeast, nine to ten days in Northeast, Midwest, and the Plains states, and sixteen to nineteen days in the Northwest and Southwest.[26] These changes have an impact on what farmers and gardeners can grow and on insect pests and diseases affecting not just agriculture but also forests, cities, and even humans.

Although a longer growing season might provide opportunities for growing crops that were previously limited by colder temperatures, such opportunities may be constrained by drought, flooding, or soils that are unsuitable for the new crops. Further, moving production zones comes at great expense to physical, economic, and social infrastructure, and can lead to conflicts as formerly productive populated areas become unproductive because of drought or heat stress.

Human Safety, Health, and Well-Being Are Threatened

Taken together, the changes brought about by climate change threaten human safety, health, and well-being. Floods pose a direct risk of drowning, and heat waves can kill vulnerable individuals like the elderly, especially those without a social support network.[27] Wildfires and dust storms during droughts impact air quality, and populations of disease-carrying organisms like mosquitoes and ticks are up, leading to possible increases in malaria, dengue fever, and other diseases.[28] Parents assessing these risks may direct their children to spend more time indoors, depriving children and families of the multiple health benefits of spending time in nature.[29] And as many environmental educators are aware, the looming threats brought about by climate change can cause stress, sadness, and related mental health issues.[30]

Addressing Climate Change

Addressing climate change involves both mitigation, principally by reducing the amount of greenhouse gases we emit into the atmosphere, and adaptation, or adjusting to the changes brought about by climate change. Consider a ski resort. It can install solar panels to power its lifts, thus helping to mitigate climate change. The resort can also adapt to warmer weather by making more snow. Whereas making more snow results in greater energy consumption and thus does not mitigate greenhouse gas emissions, some types of adaptation, most notably ecosystem-based adaption, integrate action to improve environmental quality.[31] For example, planting trees and other plants that absorb CO_2 helps to mitigate climate change. Trees and bioswale gardens also retain water and soil that otherwise would run off into rivers, thus helping communities adapt to more frequent heavy storms. Two broad strategies for mitigating climate change are (1) reducing greenhouse gas emissions (for example, by converting from coal to solar for electricity generation), and (2) increasing sequestration of CO_2 that has already been emitted (for example, by planting trees).

While adaptation is important to help ensure our short-term survival, mitigation is critical to the long-term continuation of human civilization as we know it beyond about 2050. Absent mitigation, it is estimated that prior to 2100, the earth's average global surface temperature could exceed 4°C (7.2°F) above the preindustrial average. A 2012 study conducted for the World Bank concluded that there is "no certainty that adaptation to a 4°C world is possible," and that "4°C warming simply must not be allowed to occur."[32]

Reducing all sources of greenhouse gases is important. Below we start with what an individual can do in his or her own home or school; such individual behavior change has traditionally been the focus of environmental education. Next, we talk about what people can do working together in their communities. We cannot mitigate or adapt to climate change without collective action, and thus environmental educators need to expand their efforts to get students, families, and neighbors working together to address climate change.[33] Further, environmental educators can help people influence business and government policies.

Individual and Household Behaviors

In deciding how to reduce greenhouse gases, it is important to consider the sectors—electricity production, transportation, industry, commercial and residential, and agriculture—that contribute the most emissions. Burning fossil fuels in electricity production accounts for 29 percent of U.S. greenhouse gas emissions (figure 1.3). Thirty-three percent of that electricity is consumed in homes and businesses.[34] Thus, the first question one might ask is, "How can I reduce electricity use in my home and at work?" The U.S. Environmental Protection Agency (EPA) has a number of recommendations, including purchasing Energy Star appliances and paying attention to your heating and cooling system, which can use up to half of a home's energy consumption.[35] In many counties and cities, a university extension or other program provides guidance on ways to avoid heat loss in your home and to install rooftop and community solar arrays, heat pumps, and smart meters that enable the consumer to monitor energy consumption and shift energy use to off-peak hours.[36] Nonprofit organizations, government offices, and engineering firms also can advise about financing options, including government incentive programs.

After electricity production, transportation is the second-largest emitter of greenhouse gases in the United States, responsible for 27 percent of total greenhouse gas emissions.[37]Although the solutions are obvious—walk, bike, take public transport, and reduce driving and flying—implementing them can be difficult to fit into one's life. But that doesn't mean we shouldn't try, and there can be side benefits to our health when we walk and bike. An environmental educator in Austin, Texas, worked with his son's school to develop a bike-to-school program, and U.S. cities are expanding bike-share programs based on China's model of "dockless" smart bikes.[38]

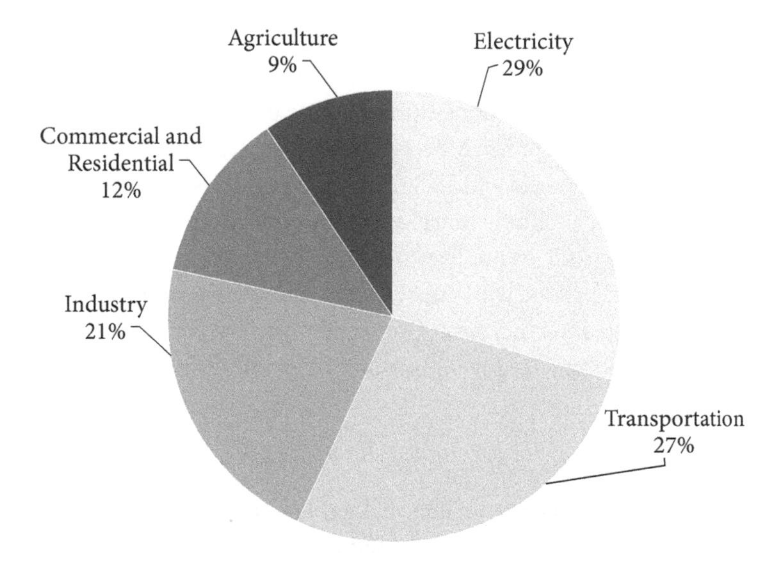

FIGURE 1.3 Sources of greenhouse gas emissions
U.S. Environmental Protection Agency, 2017

Industry is the third-largest U.S. emitter (21 percent of total emissions).[39] Considering climate change ramifications when making consumer choices can help to reduce this source of emissions. In China, the Ant Forest online game rewards consumers who purchase climate-friendly goods with points that are used to plant trees. This has the potential to reduce greenhouse gas emission by encouraging "green" choices and to sequester CO_2 through tree planting. Whereas Ant Forest uses external incentives,[40] we also make choices based on social norms and individual values, both of which can shift as we observe more and more people making green choices.[41] Educators can play a role in changing social norms by modeling and encouraging climate friendly behaviors in their programs.

Commercial and residential sectors accounted for 12 percent of total emissions in 2015. This percentage accounts for both direct and indirect emissions. *Direct* emissions result from a number of commercial and residential activities. Heating and cooling homes and businesses using fossil fuels releases CO_2. Refrigeration and air conditioning release fluorinated gases. Even waste at the landfill releases methane as it decomposes, and wastewater treatment emits methane and nitrous oxide. When we turn on the lights and use electricity produced by a power plant that burns fossil fuels to make that electricity, we are releasing greenhouse gases *indirectly*. In addition to recommending individual actions like reducing waste and turning off lights to save energy, some environmental education organizations have taken a different approach. Mass Audubon partnered with Massachusetts Energy Consumer Alliance for a "Make the Switch" campaign that promotes switching to renewable energy; their goal is for at least one thousand Mass Audubon members to switch to renewable electricity sources within a year of the campaign's start. This type of partnership may increase the likelihood that consumers will switch their energy source because the information comes from a trusted conservation source like Mass Audubon. Finally, the EPA states that agriculture contributes 9 percent of U.S. greenhouse gas emissions, much of it from livestock production.[42] Actual contributions of agriculture to greenhouse gases may be higher; some sources put livestock alone as contributing up to 18 percent of global greenhouse gases, with cows, sheep, and goats producing more emissions relative to pigs and chickens.[43] Consumers can make choices to limit meat and dairy consumption to reduce this source of greenhouse gases.

Collective Action

We have seen how addressing greenhouse gas emissions at the individual level involves consumer and lifestyle choices. But environmental education programs also engage participants in collective action and even in influencing policies. One form of collective action is scaling up individual actions, as we saw in the Mass

Audubon example—the more households that reduce their energy use or the more individuals who walk, ride their bike, or take the bus to work, the greater the impact. Other forms of collective action involve community members working together to address structural issues, such as the cost of rooftop solar installation, the lack of bike lanes, or policies that act against energy saving, which prevent people who would otherwise make green choices from doing so. In some states, citizens can work with a solar company to implement community solar, thus enabling more households to buy into renewable energy. Working with private companies to develop a car-sharing program will allow more individuals to reduce car use. Working with farmers to create community supported agriculture (CSA, or group purchasing of local produce) can reduce the need to buy packaged food; in cities where people walk or use public transport to pick up their produce, these practices also reduce gas emissions associated with transporting food.

At the local policy level, environmental education participants could advocate at town hall meetings for wind energy, bike trails, and sidewalks, and preserving forests and wetlands that absorb CO_2. They can also work with their churches, sports clubs, and other civil society organizations to implement organizational practices that reduce greenhouse emissions, such as banning single-use plastic water bottles. They may be able to help draft and implement town climate change mitigation and adaptation plans, such as New York State's Climate Smart Communities,[44] which in turn can serve as examples for other towns and spur action at the state or even national level. Environmental education participants also can call their political representatives and work for candidates who support legislation to address climate change.

Bottom Line for Educators

Like any field of science, climate science is never settled or beyond further modification. However, there is a point at which a scientific consensus is reached based on strong evidence from multiple lines of inquiry. The scientific conclusion that human greenhouse gas emissions and other activities have changed the earth's atmosphere with measurable impacts on global climate has reached that level of certainty. Moreover, climate science allows us to estimate how actions we take now and in the near future can reduce the severity of climate change in the coming decades.

Evidence of warming comes not just from climate models but from actual observations of surface, air, and water temperatures; ocean chemistry; and melting Arctic and glacial ice. In fact, much of the climate "denying" that we see is more a function of people's social and political leanings than of the facts (see chapter 5).

Disinformation campaigns by individuals and organizations who do not wish to see effective action taken to reduce climate change are an unfortunate reality. We cannot allow distortion, bias, and fabrication to prevent the evidence-based decisions and actions required at the individual and societal level to reduce climate change. The very survival of human civilization requires such action. The alternative goes beyond factual disagreement. To ignore clear evidence and fail to act, creating great peril for the near- and long-term future, is beyond a scientific, technological, or political issue; it is a question of morality, ethics, sanity, and self-preservation. Fortunately, we already have many of the scientific and technological capabilities to reduce climate change risk. We need to develop the moral compass and social and political will to use them wisely.

Environmental education can influence participant behaviors and actions at levels ranging from individual choices to local collective action to advocacy for national or global policies, and across the consumer, transport, industry, and agricultural sectors. But first we need to understand the best ways to communicate climate change and inspire action. In the next chapter we turn to explanations for varying views on climate change.

2

CLIMATE CHANGE ATTITUDES AND KNOWLEDGE

Understanding climate change attitudes is one of the challenges educators face when teaching about climate change. Attitudes are cognitive representations that summarize people's evaluation of an action, event, idea, or thing, or what social scientists call an "attitude object." In this case, the attitude object is climate change.[1]

The relationship between attitudes and behavior is not always straightforward. One might think positive environmental attitudes would engender pro-environmental behavior that minimizes environmental impacts and has positive environmental outcomes.[2] But in reality, attitudes are often a weaker predictor of behavior than we might expect.[3] In the case of climate change, although people who hold more positive attitudes toward renewable energy may be more likely to install solar panels on their home, there are many reasons why people who feel positive about renewable energy may *not* do so—for example, lack of knowledge, structural barriers such as cost, or how they feel others may view them. Generally speaking, attitudes are a better predictor of behaviors when the attitudes are more specific—for instance, if we want to predict who will install solar panels, attitudes toward renewable energy, specifically, are likely to be a better predictor than general environmental attitudes.[4] The predictive strength of attitudes also depends on whether behavioral intentions or actual behaviors are the intended outcome. In general, there is a strong relationship between believing in anthropogenic climate change and intentions to participate in pro-environmental behavior; however, the relationship between climate change belief and actual behavior is weaker.

In a study published in 2014, researchers estimated that 60 percent of adults worldwide were aware of climate change, whereas 40 percent had never heard of it.[5] Survey data demonstrate that high-emitting countries like the United States and China are among the least concerned about climate change, whereas lower-emitting countries in South America and Africa are most concerned.[6] Within countries, awareness and risk perceptions also vary markedly. A study in Cebu, Philippines, published in 2012 found that only 18 percent of fishermen in the region were aware of climate change, compared to 71 percent of laborers. This difference underlines the challenges confronting this archipelagic nation as it faces significant risks from sea level rise and ocean warming.[7]

Social scientists examining U.S. climate change attitudes over the past decade have found that those attitudes have remained remarkably stable, although acknowledgment that climate change is happening has increased steadily since 2015.[8] As of 2017, the majority—71 percent—of the country thought the climate is changing.[9] According to survey data published in 2011, 54 percent of Americans also believe that climate change is anthropogenic; but the population differs markedly in its policy preferences and behaviors.[10] At the ends of the spectrum, some Americans are very alarmed, while others dismiss climate change almost completely. Most Americans lie somewhere in between (figure 2.1). Political ideology and party identification are strong predictors of climate change attitudes and beliefs in national surveys; Democrats and liberals reliably express more alarm compared to Republicans and conservatives, who are more likely to be dismissive.[11]

Belief that climate change is happening does not equate to understanding the facts. Most Americans have heard of the greenhouse effect (87 percent,

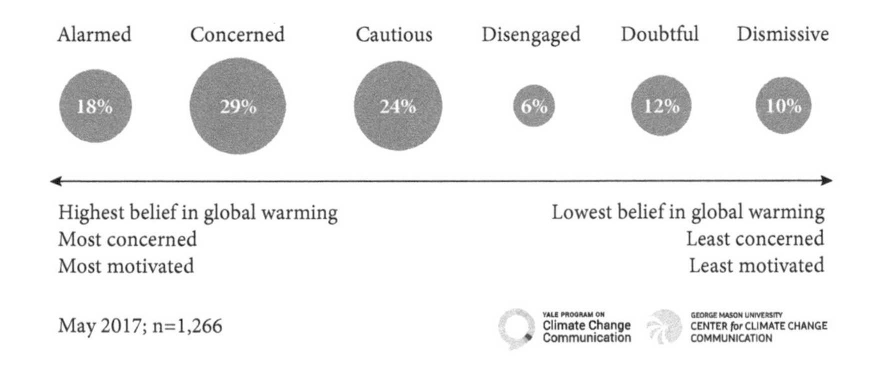

FIGURE 2.1 Global warming's six Americas
Yale Program on Climate Change Communication and George Mason University Center for Climate Change Communication, 2017

according to a 2010 Yale study), but fewer Americans understand how the greenhouse effect works, and many continue to conflate ozone layer depletion with climate change.[12] While people generally understand that carbon dioxide is a greenhouse gas that contributes to warming, they are relatively unaware of other important greenhouse gases, like methane.[13] Fewer than half the science teachers in a 2016 study could correctly identify the percentage of scientists (97 percent) who agree that humans are causing climate change, and a third reported purposefully giving students mixed messages about climate change.[14] In a study examining climate change knowledge across the United States, Canada, Germany, Switzerland, China, and the United Kingdom, higher levels of knowledge about the causes of climate change (but not the physical characteristics of climate change) were associated with higher levels of climate change concern.[15] Understanding and communicating the scientific consensus around climate change could function as a key factor in moving audiences toward supporting climate policy.[16]

Other research has focused on youth knowledge and attitudes toward climate change. A 2010 survey found that U.S. teens knew about the same or a little less about climate change compared to U.S. adults.[17] Although fewer teens said that climate change was happening compared to adults (54 percent of teens versus 63 percent of adults), more teens than adults understood that the greenhouse effect refers to gases in the atmosphere that trap heat (77 percent of teens compared to 66 percent of adults). Strong predictors for climate change concern among teens include acceptance of anthropogenic climate change, frequency of discussion with family and friends, and the perceived acceptance of anthropogenic climate change by family and friends.[18] The importance of family and friends suggests intergenerational programs that engage parents alongside children could be particularly effective at building concern. Efforts like the Climate Urban Systems Partnership in Pittsburgh employ this approach; when families come to the program's booths at festivals, the children and parents explore climate change activity kits together.[19]

Because much of the research on climate change attitudes and knowledge comes from Western countries, "debates remain anchored primarily in the experiences, values, and desires of developed nations . . . even when we think we are arguing against what we construe to be the selfish interests of 'the West.'"[20] Educators working in other countries may find themselves confronted by challenges different from those facing U.S. educators. For example, whereas political polarization dominates climate change discussions in the United States, this is not true in other countries that are major players in global climate negotiations, like India.[21]

Bottom Line for Educators

Understanding audiences' climate change attitudes and knowledge can guide educators in developing program outcomes and content. For example, if educators work with audiences who are already very concerned and knowledgeable about climate change, they may want to focus on developing audiences' sense of collective efficacy—the feeling that they can respond collectively to the problem—to help them avoid despair. Educators working with audiences who are more skeptical or less aware of the problem may target climate change knowledge and awareness as preliminary outcomes. These educators may also seek to identify areas of common ground, such as shared experiences in their communities, that enable them to have a positive dialogue with audiences even as they disagree about certain aspects of climate change. It may hearten environmental educators to look at the opinion data and remind themselves that the majority of Americans *do* believe that the climate is changing, even if they disagree on ways to address the problem.

3

CLIMATE CHANGE EDUCATION OUTCOMES

We often talk about our environmental education program's success, but we may fail to specify the outcome that defines that success. Are we successful at instilling climate literacy? At fostering self-efficacy and individual pro-environmental behaviors, or at sparking deliberation, civic engagement, and collective action? Or maybe our intended outcome relates to positive youth development, inspiring hope, or some type of resilience?

Defining program outcomes is the first step in a program development cycle and works in concert with educators' audience assessments. Climate change education outcomes are any desired changes that result from climate change education programs and that are intended to enhance natural and human systems by reducing greenhouse gasses and, where needed, helping communities adapt to climate change in an environmentally sound manner.[1] Specifying outcomes enables educators to strategically plan and evaluate program activities. It is important to note that making a program's desired outcomes transparent also enables educators to reflect on how likely their activities are to reach their goals, and to adjust if necessary. Below, we describe three categories of climate change education outcomes drawn from environmental education and climate change communication research, as well as from our own experience and conversations with climate change educators: individual outcomes like climate literacy, attitudes, self-efficacy, and behavior change; community outcomes like collective efficacy, social capital, and collective action; and direct environmental outcomes. We conclude with a short discussion of resilience, a term that can refer to individuals, communities, ecosystems, and integrated social-ecological systems.

Climate Change and Environmental Education Outcomes Focused on Individuals

Climate literacy (knowledge and skills), positive attitudes and emotions, and self-efficacy all contribute to changes in individual environmental behaviors.

Climate Literacy

Climate literacy blends knowledge and skills. *Climate knowledge* includes knowledge about climate systems and processes, and about how humans affect climate, how climate change affects humans, and what actions humans can take to mitigate and adapt to climate change. While knowledge is one factor in people's decision to act pro-environmentally (or toward climate change solutions), knowledge on its own is not sufficient for motivating behavior change.[2] Knowledge differs from *climate awareness*, which is a broad concept referring to knowing that something exists.[3] In environmental education, awareness refers to an individual's perception of, influence on, and concern for the environment.[4] Climate-related skills are abilities that enable someone to perform a task and lead to a desired action or goal over time.[5] Climate literacy skills include communicating about climate change, assessing climate-related information,[6] and participating in constructive dialogue.

According to *Climate Literacy: The Essential Principles of Climate Science*, a climate literate person

- understands the essential principles of the earth's climate system,
- knows how to assess scientifically credible information about climate
- communicates about climate and climate change in a meaningful way, and
- is able to make informed and responsible decisions with regard to actions that may affect climate.[7]

The National Academy of Sciences also includes behavior change under the umbrella of climate literacy.[8]

Environmental literacy is broader than climate literacy. Environmental literacy is defined as the ability to make informed decisions concerning the environment; being willing to act on these decisions to improve the well-being of other individuals, societies, and the global environment; and participating in civic life.[9]

Attitudes and Emotions

Attitudes and emotions are included in definitions of environmental literacy but are generally left out of definitions of climate literacy. However, understanding attitudes and emotions is critical to designing climate education programs.

Attitudes are cognitive representations of how people evaluate an action, event, idea, or thing.[10] The building blocks for attitudes include values, beliefs, and emotions.[11] Many environmental education programs operate with the goal of promoting positive attitudes toward the environment, although research shows that attitudes are very hard to change.[12] A large body of climate change communication research focuses on understanding public attitudes toward climate change and on experimenting with the ways in which communication affects attitudes. This research can help environmental educators target their message while avoiding trying to directly change attitudes.

ATTITUDE COMPONENTS

Educators may seek to understand attitudes broadly or to target the specific components of attitudes.

- *Values*, the guiding principles by which people live and preferences about how society should function,[13] form the basis of attitudes.
- *Beliefs* represent a person's subjective knowledge about the world; beliefs are distinguished from knowledge, which is based on facts.[14] For example, in the climate change context, one person might hold the belief that the planet is experiencing a pronounced warming period because of anthropogenic emissions of greenhouse gases, whereas another may believe that the warming trend simply reflects natural variability in global temperatures. A third person may believe that climate change is a hoax and discount the temperature evidence entirely. These different beliefs about the same knowledge (warmer temperatures) hold implications for attitudes about climate change and behavioral intentions.
- *Emotions* are related to beliefs and attitudes. For example, if a person believes that climate change poses a serious threat to human society, including a direct personal threat, that person may experience the emotion of fear and demonstrate fearful attitudes. While fearful attitudes can be pro-environmental attitudes and foster pro-environmental behaviors, fearing climate change might also result in a terror management response that actually decreases the likelihood someone will engage in pro-environmental behavior (see chapter 6).

EMOTIONS: HOPE AND FEAR

Emotions not only play a role in defining attitudes; they also can be education outcomes. Emotions are psychological and physiological responses to stimuli, and they guide information processing and behavior.[15]

Hope is an emotion that consists of goals (what we want to happen), pathway thinking (our ability to figure out how to meet those goals), and agency thinking (a motivation to use those pathways).[16] Hope has been associated with willingness to engage in pro-environmental behavior.[17]

Fear is an uncomfortable emotional response to a perceived threat.[18] Although the media frequently employ fear appeals to communicate about climate change, a more effective strategy may be to inspire hope in audiences with the goal of supporting climate change action.

Self-Efficacy

Self-efficacy refers to confidence in your ability to achieve goals. It is an "empowerment variable" and an important factor in fostering responsible environmental behavior.[19] A strong sense of self-efficacy results in people expending more effort in the face of obstacles like climate change.[20]

Environmental Behaviors

Environmental behaviors are changes made by an individual, such as installing solar panels on one's home, turning down the heat, reducing consumption, or purchasing a fuel-efficient vehicle.

Climate Change and Environmental Education Outcomes Focused on Communities

Positive youth development, in which youth develop leadership, communication, and other assets, is included in this section on community-level outcomes. This is because youth assets are often an outcome of environmental education programs that involve community gardening or some other form of collective action; youth with leadership skills and other assets are also more likely to engage in collective action.

Youth Assets

Positive youth development is an asset-based approach to promoting young people's well-being physically (e.g., through good health habits), intellectually (e.g., through critical thinking), psychologically (e.g., through building confidence), and socially (e.g., through trusting others).[21] In addition to helping youth live productive lives, such assets help enable youth to engage in positive environmental behaviors and collective actions. Environmental education—in particular programs where youth build citizenship skills through civic engagement—can be a way to achieve positive youth development.[22]

Social Capital

Social capital has multiple definitions focusing on the personal and collective or social benefits of connections with others. It generally refers to the relationships, trust, and shared norms that individuals can draw on to work through collective problems like allocating green space or taking action to mitigate climate change.[23]

- *Trust* is a key component of social capital and a prerequisite for collective action and collaboration. Researchers have defined trust in multiple ways, but these definitions all converge around the idea that trust is a psychological state that accepts some form of vulnerability and concerns the person who trusts, the person who is deemed trustworthy, and an action.[24] Environmental education programs in which youth work together to achieve a challenging goal can build trust among participants.[25] In climate change communication, one important aspect of gaining trust is the use of a trusted messenger (see chapter 9).[26]
- *Positive dialogue* can build the connections that are part of social capital.[27] Climate change education programs that promote discussions requiring participants to reflect on the trade-offs of their own views while also acknowledging the benefits of fellow participants' views may be particularly effective at fostering positive dialogue.[28]

Collective Efficacy

Collective efficacy is a belief that a group can achieve its goals.[29] Audiences can easily feel "disempowered by the scale of environmental problems" and can benefit from "opportunities to work for social and environmental change with others to acquire a collective sense of competence."[30] Like trust, collective efficacy is a stepping-stone to collective action.[31]

Collective Action

Collective action is action taken by a group in pursuit of perceived shared interests.[32] Examples of collective actions to address climate change include volunteer groups restoring dune habitat to enhance shoreline stability, or a neighborhood forming a renewable energy collective or working with a city or town to improve infrastructure for bike commuters. Similar to environmental action, climate change action integrates science knowledge and civic engagement, and can also include voting and other policy-related activities.[33]

Climate Change and Environmental Education Outcomes Focused on the Environment

Positive environmental outcomes are direct environmental improvements such as restoring ecosystem services or improving water and air quality. In climate change education, mitigation and ecosystem-based adaptation are environmental outcomes.

Climate Change Mitigation

Climate change mitigation is defined as slowing down climate change by reducing and stabilizing greenhouse gas emissions or enhancing greenhouse gas sinks.[34] Climate mitigation actions include planting trees and switching to renewable energy to reduce use of coal, oil, and gas.

Climate Change Adaptation

Climate change adaptation is about helping people and environments prepare for and adjust to climate change. Adaptation efforts attempt to reduce society's vulnerability to climate change impacts, like sea level rise, extreme storm events, and water and food shortages, and include activities like installing permeable pavement to reduce flooding or restoring dunes to slow shoreline erosion.[35] Note that not all adaptation practices are consistent with environmental education principles. New coastal developments may include houses raised on stilts, reducing the likelihood of the homes flooding but failing to address the underlying environmental issues. Environmental education can play a role in helping adaptation efforts incorporate mitigation outcomes.[36] Ecosystem-based adaptation activities, like restoring and maintaining mangroves, mitigate climate change by ensuring the mangroves remain a carbon sink while helping communities adapt to climate change by reducing flood impacts related to sea level rise.

Resilience

Resilience is a term we hear frequently in relation to climate change (table 3.1). In general, resilience refers to the ability of individuals, communities, ecosystems,

TABLE 3.1 Resilience definitions

TYPE OF RESILIENCE	DEFINITION
Psychological resilience	The processes of, capacity for, or patterns of positive adaptation during or following exposure to adverse experiences that have the potential to disrupt or destroy the successful functioning or development of the person[38]
Community resilience	The ability of communities to cope with and recover from external stressors resulting from social, political, and environmental change[39]
Ecological resilience	The magnitude of disturbance that a system can experience before it moves into a different state with different controls on structure and function[40]
Social-ecological systems resilience	The capacity of social-ecological systems to continually change, adapt, or transform to maintain ongoing processes in response to gradual and small-scale change, or transform in the face of devastating change[41]

and social-ecological systems to respond to change, including hardship and disasters. Climate change educators working in areas impacted by major storms, drought, wildfire, or other natural disasters exacerbated by climate change may be particularly interested in fostering community and social-ecological resilience.[37]

Bottom Line for Educators

Climate change education programs target a variety of environmental education outcomes that overlap with climate change communication goals. Defining outcomes early on in the program development process will assist educators in choosing appropriate activities to meet their goals and their audiences' needs.

4

CLIMATE CHANGE EDUCATION VIGNETTES

Given what we now know about climate change perceptions and potential education outcomes, how might one approach climate change education with different audiences, intended outcomes, and educational settings? One of the authors of this book, Anne Armstrong, has worked as an educator in a variety of settings, from national and state parks to nature centers and residential facilities. From 2010 to 2015 she also worked as the education director for the Chincoteague Bay Field Station, a nonprofit environmental education and research field station in Wallops Island, Virginia. There, she and the education staff juggled the needs and interests of multiple audiences: teachers and students attending multiday coastal ecology field trips; university faculty and students conducting research, teaching, or attending courses; and children and adults of all ages who came for overnight educational experiences. Based on Anne's graduate research on environmental educators' use of climate change communication research and her experience as an environmental educator, in this chapter we introduce three fictional climate change education scenarios: Elena leads a community meeting to plan a collective action; Jayla works at a nature center; and Will teaches in a high school classroom. We weave these vignettes into the subsequent chapters and discuss connections between research and the three educational approaches.

Vignette 1: Community Climate Change Conversation—Elena

Elena works for an environmental education organization about to embark on a living shoreline project. Living shorelines use oyster reefs, marshes, and other existing and restored habitats to manage coastal erosion.[1] Elena's organization is trying to build local support for action on climate change mitigation and ecosystem-based adaptation. She hosts a community meeting at the local volunteer fire department to introduce the organization's plans and to recruit volunteers for a marsh planting and artificial oyster reef ("oyster castle") installation the following month. Elena starts by asking community members to share some of the changes they've seen in their community over their lifetimes.

John raises his hand and explains that where his father had grown strawberries is now salt marsh. Cindy describes how the pine tump (a small mound with pine trees growing on it) has disappeared over the last five years from the mouth of Old Pine Creek. Jim says his house now floods regularly during spring tides but didn't used to when his grandfather owned the house.

Elena says, "We all want to protect this area, and it's going to take us working together collectively to do so. All of the changes you've listed are a result of sea level rise caused by warming temperatures—climate change."

John raises his hand and says, "Now I'm not sure about this climate change. It sounds like a bunch of hooey to me, and from what I've seen has been much too overblown by the media and liberal politicians. I know that my land is changing, but I can't attribute that to global warming. I've had to shovel more snow over the past two years than I have my whole life put together."

Elena responds, "Thanks for sharing, John. We had a cold, snowy winter here, but overall, our annual average temperatures have been rising, as have ocean temperatures. Mining and burning fossil fuels like coal, oil, and natural gas releases carbon dioxide and methane into the atmosphere. The CO_2 and methane act like a big heat-trapping blanket, warming the air and the ocean. These warmer temperatures contribute to sea level rise in two ways. First, warmer temperatures melt glaciers and ice caps on the land, and the meltwater makes sea level rise. Second, warm ocean temperatures actually cause the water to expand. As the water expands, we get increased flooding and erosion."

"Well, I don't know about all that. Like I said, I shoveled more snow in the last two years than ever in my life. Also, if the ocean's getting warmer and we start having more warm-water fish up here, I'll get my shrimping license!" (At this time, the town is too far north for shrimp populations large enough to support an industry.) The audience laughs and nods along with John.

Elena's face gets a little red, but she decides that instead of continuing to debate with John, she'll approach the project from a different perspective. "OK, OK," she says, waiting for the audience to quiet down. "Clearly, we'll agree to disagree for now on the causes of coastal erosion. But here's what I think we agree on. As you've seen, the beach has experienced a lot of erosion since Hurricanes Irene and Sandy. This town used to be an oyster capital. We'd like to take steps toward reclaiming that title. We'd like to start building the beach back, using native marsh plants and constructing an artificial oyster reef using oyster castles just offshore that will support oyster populations while slowing wave energy. We can't do it on our own, though, and will need all the help we can get."

Alice raises her hand with a concerned look on her face and asks whether the shoreline restoration project will mean town members are no longer allowed to go clamming on the beach. Someone else asks whether the state marine resource commission and the U.S. Army Corps of Engineers have cleared the project, because "those oyster castles might be navigation hazards." Elena explains that they don't want to deny anyone access to the site, even if it is her organization's private property, and assures the group that she has gone through the proper permitting channels. Finally, John raises his hand and says, "Well, I like the idea of building the beach back and bringing the oysters back, but I'm still not convinced that our problems are due to climate change and not just plain old erosion. When you have the president of the United States call Tangier Island to tell them that *they're* going to be all right, I just have to wonder whether we're making a fuss over nothing."[2] Elena sees some people in the audience nodding their heads and looking at her expectantly, waiting for her response.

Elena takes a deep breath and says, "Thanks for your input, John. I'm glad that we can agree on building the beach back." Elena goes on to describe the plans for the shoreline and puts out a volunteer request. The next month, she has twenty volunteers, including John, from the town shoveling sand and oyster shells, and planting native marsh grasses.

TABLE 4.1 Elena's vignette: setting, outcomes, audience

Setting	Community gathering in a rural coastal community. An environmental education organization is recruiting volunteers for a living shoreline project in a community suffering from recurrent flooding because of sea level rise.
Outcomes	Ecosystem-based adaptation Collective action Climate literacy
Audience	Adult members of the community in which the educator, Elena, hopes to construct the living shoreline Attitudes: Some are cautious and concerned, while others are dismissive. Knowledge and beliefs: The audience knows that the land is changing and eroding but is not fully convinced of links between the flooding that causes the erosion, sea level rise, and climate change.

Vignette 2: Nature Center Interpretation—Jayla

Jayla has funding to make new exhibit panels along the popular Monkey Flower Meander trail at the Chaparral Nature Center in a large California city. According to a recent survey her center conducted, visitors are concerned about climate change and want more information about how it will impact their region, particularly regarding water resources. They also want information about actions they can take. Jayla decides one way she can address climate change at the nature center is by incorporating a climate change message related to water at one of the posts on Monkey Flower Meander. Jayla starts by doing some online research about climate change in general and in California. She pulls together information from the U.S. National Oceanic and Atmospheric Administration, the Environmental Protection Agency, and the National Aeronautics and Space Administration and formulates potential messages that are locally relevant and tie into the plants highlighted on the interpretive trail.

She organizes a focus group of six of the center's most dedicated volunteers. She starts by asking them to talk about climate change so she can get a baseline idea of their understanding of the phenomenon. She then asks them to think of potential solutions to their city's water availability issues and creates a list from those ideas. She shows them her draft exhibit panel text (figure 4.1), and they give her feedback on what they like and dislike. Below are some comments volunteers give Jayla.

What they like:

- Using the popular trail to teach about climate change
- The potential to tell climate stories about plant species represented on the trail
- Linking climate change, water issues, and native plants

What they think could be improved:

- A better explanation of why visitors should care about chaparral plants shifting their ranges. They suggest tying chaparral to visitors' sense of place by framing chaparral as a defining feature of the local landscape.
- Tighter connections to center resources and to climate action. Volunteers think the actions Jayla provides are too broad and do not touch directly on temperature and water issues in their city. They suggest mentioning the center's rain barrel construction program as an easy entry point for water conservation in private homes, and the city's ample bike lanes and ride-share program that can reduce greenhouse gas emissions.

Climate Change in the Chaparral

Chaparral plants are already changing their ranges in response to climate change[1]

Climate change is caused by an increased concentration in greenhouse gases in the atmosphere due primarily to burning of fossil fuels. Carbon Dioxide, methane, and nitrous oxide form a heat-trapping blanket around the earth; the more of these gases we add, the warmer our planet will become.

In California, the increased temperatures due to climate change will increase the amount of evaporation that occurs from soil and plants, reducing the amount of available water for human communities and natural communities alike.

Biologists have already seen changes in California's plant communities due to rising temperatures. Plants like this creosote bush you see on the monkey flower meander trail have shifted ranges are now found at higher elevations than they were 20 years ago. Higher elevations are cooler and experience less evapotranspiration, enabling plants to hold on to more water.

We can accomplish a lot when we work together!

Together, we can reduce California's greenhouse gas emissions and conserve precious water supplies.

Shopping at your local farmer's market shortens the supply chain from farmer to plate and reduces the amount of carbon dioxide emissions necessary for getting food onto your table. You get fresh food and you support local industry! Support local renewable energy initiatives that make wind and solar more affordable and accessible to a wider population. Removing nonnative grasses and using native vegetation in your yard saves precious water and makes your property less vulnerable to wildfires.

[1]Kelly, Anne E, and Michael L. Goulden. 2008. "Rapid shifts in plant distribution with recent climate change." *PNAS* 105 (33): 11823–11826.

FIGURE 4.1 Jayla's draft version of a climate change exhibit panel

TABLE 4.2 Jayla's vignette: setting, outcomes, audience

Setting	Urban nature center in Southern California. A nature center interpreter integrates climate change topics into an interpretive exhibit panel.
Outcomes	Climate change knowledge about individual behaviors to mitigate climate change
Audience	Adult visitors Attitudes: Most are adult visitors concerned about climate change. Knowledge: Adult visitors know that the climate is changing and want more information about impacts to their region and actions they can take.

Vignette 3: High School Classroom—Will

Will is a science teacher in central Kentucky. Although his specialty is biology (he really loves entomology), he finds himself also teaching earth science, which includes a section on climate change. When the Next Generation Science Standards were adopted by Kentucky,[3] Will and many nearby science teachers realized they did not have the background to effectively teach the climate change components. Will is wary of the first day of class. He knows his students aren't necessarily aware of their various political leanings yet, but their parents definitely are. Will does not think Kentucky, a coal state, is the most receptive place for climate change education.

Will's plan is to depoliticize his teaching as much as possible and to teach the topic in a way that makes students hopeful about the roles they can play in their community. Will recently completed a Project Learning Tree "Southeastern Forests and Climate Change"[4] training module from the Kentucky Association of Environmental Education, and he's planning on starting with an activity called "Clearing the Air," which gives students a chance to examine common climate change misconceptions and weigh them against scientific evidence. Most of his students enjoy the exercise, although a couple sulk through the presentation on climate change evidence. Will also uses the Kentucky supplement for the activity to connect the trees students see regularly to climate change impacts. He decides to test out an analogy he read on *National Geographic*'s website about CO_2 emissions and the atmosphere. "Imagine the atmosphere is a big bathtub," he tells students, "and trees and soil are two of nature's ways of draining that bathtub. As long as we pour CO_2 into the atmosphere faster than nature drains it out with trees and soil, our planet continues to warm."[5]

In the second week of the course, Will asks students to brainstorm ideas for a climate change mitigation project that they can pursue as a class. They want to plant trees on their school grounds and then monitor how much carbon the trees sequester. Will gets permission from the school and then starts brainstorming

TABLE 4.3 Will's vignette: setting, outcomes, audience

Setting	High school science classroom in Kentucky
Outcomes	Climate literacy Collective action Positive youth development
Audience	High school students Attitudes: Disengaged prior to the class Knowledge: They know that climate change is an issue and that it is contentious, but do not know very much about climate science.

how to fund the project with the students. They come up with a suite of options and organize themselves into project teams. The students send out donation request letters to local nurseries, run bake sales at school, and partner with the local ice cream parlor to have 10 percent of a day's profits go toward the project. That spring, they plant fifty trees in their school yard.

Three Approaches

Elena, Jayla, and Will each consider their audience's climate change knowledge and attitudes as they formulate their program strategies, and they each hope to inspire or engage their audience in climate change action. Elena may have confused her audiences' admission of changes in the landscape with belief in climate change, and she has to adjust her expectations accordingly. Instead of pushing John further on the issue of whether climate change is real, she focuses on what they have in common to bring John onboard with her intended outcome of climate change ecosystem-based adaptation. Jayla's nature center conducted a visitor survey that Jayla uses to inform her exhibit panel—she knows that her audience is concerned about climate change, so she focuses on enhancing their knowledge about local climate impacts and potential actions they can take. Additionally, Jayla polls her volunteers and has them review her draft messages to further ensure the interpretive panel reflects her audience's needs and attitudes. Will is quite worried about his students' attitudes because of what he knows of their parents' political leanings (see chapter 5). He hopes that by having students consider each other's perspectives, he can reduce polarization in his classroom.[6] Like Elena, he involves his audience in a collective climate action, although in his case, the students develop the plan, whereas Elena approaches her audience with her plan in hand. As we proceed, we will continue exploring Elena, Jayla, and Will's approaches in light of research in psychology and climate change communication.

PART 1 RECAP

Scientists have been studying the effects of greenhouse gases in the atmosphere for well over a century. Nearly all scientists agree that human activities are causing climate change. Environmental educators, with their close ties to local communities and experience teaching about environmental issues, can influence climate change action at a range of levels, from individual household behaviors like reducing energy use to community-scale actions like advocating for bike lanes or increasing access to renewable energy options. Although much of the U.S. public is aware that climate change is happening, for many climate change is not a top concern, and opinions are polarized along political lines. Educators can choose from a wide range of outcomes when designing climate change programs. Like Elena, Jayla, and Will, they can use their knowledge of their audiences' climate change attitudes to guide program outcomes, content, and structure.

Tips for Educators

1. Learn the basics of climate change so you feel confident interpreting the science for your audience and can present options for climate action.
2. Learn about your audience's attitudes and knowledge. Jayla learns about her audience through a survey, but you can also use tools like Yale's Climate Change Opinion maps[1] to find out general opinions in your county.
3. Choose a program outcome that aligns with your strengths and is realistic, given what you know about your audience.

4. Take comfort in the knowledge that the majority of people in the United States believe the climate is changing, even if they disagree on why or what should be done. If you work with audiences who are doubtful or dismissive, consider adjusting your outcomes to reflect the realities of what you can accomplish together. Elena decides to forgo trying to persuade John about climate change, yet she still engages him in a climate change ecosystem-based adaptation action.

Part 2

THE PSYCHOLOGY OF CLIMATE CHANGE

If Elena, Jayla, and Will were interested in digging deeper to understand the origins and social contexts of their audiences' attitudes, they could look to environmental psychology research. Over the past two decades, social science researchers have explored why the issue of climate change or, indeed, environmental issues in general, fails to drive people to action. One significant finding is that a host of psychological processes can contribute to so-called "dragons of inaction"[1] that hinder individuals and groups from taking climate action. These processes are rooted in our identities and the sense of distance that many people feel from the issue. Psychology also proposes theories, such as terror management and cognitive dissonance, that help explain climate change denial. In the absence of understanding of these processes, "increasing climate education is unlikely to make a significant contribution to public consensus on the issue."[2] Further, people differ in their climate-related views depending on their social identities—politically conservative, liberal, or a particular ethnicity. Thus, climate change educators need to understand audiences' identities and worldviews to design programs to reflect the values, attitudes, and experiences audiences bring to the program.

In part 2, we explore research on how social identity affects climate change attitudes, how the distance from which people perceive climate change impacts the way they engage with climate change, and mechanisms that explain climate change denial. This research can inform strategies to achieve a suite of climate change education outcomes, from climate literacy to self- and collective efficacy and promoting positive dialogue.

5

IDENTITY

One of the authors of this book, Anne Armstrong, once participated in a two-day training session about the impacts of sea level rise on the Eastern Shore of Virginia. At the start of the training, a man raised his hand and said that he wanted to know what was causing sea level rise and that if he knew what was causing it, he would have a better idea of how to solve the problem. A local scientist explained the links between climate change and sea level rise, but the man did not accept his explanation. During the second day of the workshop, after several other presentations that connected climate change to sea level rise, the man repeated his question from the first day, insisting that if someone could just explain the cause of the problem to him, we would be better able to come to a conclusion about the best solutions. This man was a respected leader in the community, with a degree in civil engineering. He was as capable of understanding the science as Anne was, and yet they came to vastly different conclusions. What was going on?

Looking at Americans' lack of climate change understanding, it is tempting to conclude that the problem is a deficit in the public's climate change knowledge. In education and communication, this idea falls under what is called the science comprehension thesis or information deficit model[1]—the idea is that the public lacks the information scientists have, and once they have it, they will be more likely to accept human-caused climate change and support policy change and other action[2] (table 5.1). This represents a very attractive proposition for communicators and educators alike because it is so simple. Unfortunately, research suggests that this model is far too simplistic and that additional factors besides knowledge influence the decisions that audiences make.[3] Although knowledge is certainly a factor in people's decision to act pro-environmentally (or toward climate change solutions), knowledge, like attitude, is not sufficient on its own for motivating behavior change.

Identity Theory

Identity plays an important role in how people engage with climate change information. Identity is "fundamentally a way of defining, describing, and locating oneself."[4] Humans have multiple identities. Environmental educators may be familiar with environmental identity, or "a sense of connection to some part of the nonhuman natural environment that affects the way we perceive and act toward the world; a belief that the environment is important to us and an important part of who we are."[5] Environmental identity can be a type of personal identity, connected to someone's sense of who they are as an individual. It can also be a type of social identity that encompasses how people position themselves in relation to others.[6] This section will focus primarily on different types of social identities and their association with climate change attitudes and behaviors.

Social identity represents one lens through which researchers have investigated attitudes toward and engagement with climate change information. As social animals, our identity is derived in part from the groups to which we belong.[7] Group memberships range from those that are largely determined, such as citizenship or ethnicity, to those that reflect individuals' values, interests, and skills (e.g., environmental educator, birder, Democrat). The norms, or expectations, of the groups we belong to influence our attitudes, beliefs, and behaviors. Moreover, although we typically think about identities as stable, research suggests that identities are dynamic; any one of our different identities can become activated or "salient," and guide our behaviors in response to social and situational cues.[8] Environmental educators, for example, may feel pressured to use a reusable water bottle or coffee mug because they perceive this behavior as normative within the group that defines their social identity—this may be especially true in contexts that cue this identity, such as attending an environmental education conference.

Social identity can affect the way people process information. In contrast to the science comprehension thesis, which suggests that people learn and act on facts, people often interpret new information in ways that align with and reinforce their group commitments. This process is known as motivated reasoning[9] (table 5.1). Motivated reasoning affects which information people consider as they think about a given issue and how they use that information to make judgments or draw conclusions.[10] Thus, someone who is alarmed about climate change and someone who dismisses climate change can attend the same climate-related program, yet they hear different things and come away with very different conclusions, based in part on their preexisting ideas and related group commitments and social identities.[11]

Identity protective cognition is a type of motivated reasoning (table 5.1). When identity protective cognition is activated, people avoid beliefs that might alienate them from their chosen group as a means of protecting their sense of self. While denying that climate change exists might seem irrational to some people in the context of scientific consensus, it may be a perfectly rational conclusion from a social identity perspective if your peers and your group also deny climate change.[12] During Elena's meeting (see chapter 4), John seems to ignore the evidence presented by Elena based in part on his perception of alarmism from the media and liberal politicians. This is an example of identity protective cognition—John doesn't believe the facts that might alienate him from his conservative identity. In Will's class, some students remain skeptical after having participated in classroom activities, suggesting that something other than an information deficit, perhaps their social identity as members of a conservative family or church, is keeping them from engaging fully in the topic.

Motivated reasoning and identity protective cognition can each contribute to confirmation bias, in which people look for information that confirms what they already know or think, leading them to dismiss ideas that might require them to change their behavior.[13] In Elena's case, John's personal experience with cold weather confirmed his view that climate change was made up and disconfirmed Elena's explanation of climate change. In a polarized media environment rife with opportunities for selective exposure—for instance, in which conservatives watch Fox News and liberals watch MSNBC—each faction is exposed to information that is likely to confirm and bolster its preexisting beliefs.[14] In this way, viewers of Fox News and MSNBC may consider themselves just as knowledgeable about climate change, but this knowledge may lead the two groups to very different conclusions.

TABLE 5.1 Theories about how people assess climate change information

THEORY	EXPLANATION
Science comprehension thesis	Conclusions drawn based on information. Climate change information leads to climate change action.
Motivated reasoning	Conclusions drawn based on what you *want* the conclusions to be. Motivated by previous knowledge, values, and beliefs.
Identity protective cognition	A type of motivated reasoning. Conclusions drawn based on what you want the conclusions to be so as to be consistent with your peers and your social group.

Identity and Climate Change Research

Identity protective cognition related to climate change has been closely linked to political affiliation.[15] When faced with the same information about climate change, Republicans and Democrats polarize on their willingness to support climate policy.[16] One study demonstrated that Republicans with higher education were *less* likely to say they were concerned about climate change. In contrast, education was positively correlated with climate change concern among Democrats.[17] The study suggests that, in addition to providing individuals with greater knowledge about climate change, education gives people access to a wider set of information that can be used to argue for or against the problem's existence or urgency in a motivated fashion that reinforces a social identity like political affiliation. Another study compared people's science literacy with their belief in climate change and found a similar trend. For conservatives, higher levels of science literacy were *negatively* correlated with belief in climate change; for liberals, however, higher levels of science literacy were *positively* correlated with belief in climate change. When opinion surveys ask people if they "believe" in climate change, their answer may say more about their values and their social identity than their knowledge of climate science. Importantly, whereas much of the research on motivated reasoning in the context of climate change has focused on the biased interpretation of information among Republicans and conservatives compared to Democrats and liberals, recent evidence suggests that partisans on both sides are prone to engage in motivated reasoning about climate science.[18]

Will's climate change lessons triggered some students to express their identities as climate change skeptics, and they sulked through his presentation. Whereas Jayla works mostly with visitors who believe in climate change, Elena and Will are challenged to handle identity-protective cognition during their programs. Elena seeks to find common ground with her audience by drawing attention to their shared sense of town identity and heritage and by emphasizing the co-benefits associated with her shoreline project. Co-benefits are the non-climate related benefits that arise from climate adaptation and mitigation projects. A co-benefit of Elena's project is reduced erosion risk. Will assumes his students' parents have distinct political identities that would make them opposed to his teaching climate change in the classroom. Will chooses an activity that requires students to reflect on their own and their peers' attitudes toward climate change, which could assist in moderating extreme attitudes.

Although political orientation is a strong predictor of climate change belief in the United States,[19] the influence of social identity on climate change views goes beyond political affiliation.[20] Racial and ethnic identity also predicts climate change attitudes and risk perceptions, and some studies have found that members of racial and ethnic (nonwhite) minority groups in the United States report higher levels of environmental concern and support for climate change policies relative to whites. When it comes to climate change, examining the interaction of political and racial/ethnic identity has revealed a more nuanced view of the factors that predict climate change attitudes in the United States, as survey data suggest that political orientation is a weaker predictor of public opinion on climate change among nonwhite minorities than among whites. This weaker association may be attributable, in part, to different levels of experience with environmental impacts across groups, given that minorities are disproportionately affected by environmental pollution and negative climate change impacts.[21] For example, anthropologists documenting climate change adaptation priorities in two African-American communities on Maryland's Eastern Shore that are vulnerable to sea-level rise faced no political opposition to their informational sessions on climate change science.[22] The overall population in the counties in which these communities are located, however, exhibit lower-than-average rates of acceptance of anthropogenic climate change.[23] There is, in fact, a documented "white male effect" in risk communication literature whereby white males view environmental risks like climate change as less important than do white women and minorities.[24]

These results have important implications for environmental educators. They point to the complex ways in which social factors shape how people engage with climate change as a uniquely global problem, one that has asymmetric causes and impacts across groups and that will require unprecedented cooperation both within and between nations—features with potentially rich implications that psychologists are only beginning to address.[25] Programs and messages that consider climate-related injustices faced by minorities and other groups will likely fare better than those that deal with climate change impacts in the aggregate.[26] Mainstream environmental organizations lack ethnic diversity throughout their ranks and do not often collaborate with ethnic minority or low-income organizations.[27] High minority participation in community gardening and environmental justice organizations[28] points to possibilities for building stronger and more diverse coalitions around climate change.

Identity and Climate Change Education

Climate change education is certainly not immune to the effects of identity on students and teachers and is subject to outside pressure from lobbying groups. In 2017, the Heartland Institute sent tens of thousands of science teachers its book, *Why Scientists Disagree about Global Warming*, and an accompanying DVD questioning the scientific consensus on climate change.[29] Classroom teachers like

Will can find it challenging to overcome the effects of identity-based cognition in the classroom.[30] However, some good news is emerging from research on climate change education and identity.

Although past research on identity and environmental education centered primarily on environmental identity,[31] a new body of research exploring social identity and climate change is emerging. A study in North Carolina revealed that high school students are less influenced by political polarization than adults are, and that increased climate knowledge correlated with increased acceptance of anthropogenic climate change regardless of political leanings. The researchers suggest that one reason for this difference is that high school students may hold less entrenched worldviews and values than adults.[32] Researchers have also found that identity does not appear to drive the perception of climate change risk for nonhuman life. High school students in one study relied on mental shortcuts, or heuristics, to assess climate change risk to human society, and these shortcuts were determined in part by their political affiliation. When they assessed risk to wildlife, however, they relied on their knowledge of climate change.[33]

Bottom Line for Educators

Acknowledging the role of identity in climate change attitudes and behaviors can help environmental educators in program planning, including developing suitable outcomes and tailoring messages for particular audiences. Environmental educators can key into particular social identities and connect climate change to these identities. They also may focus on "superordinate" identities, such as "coastal Virginian," that cross political identity,[34] and choose outcomes, like replanting marshes, that appeal to all regardless of political identity. Jayla's draft exhibit may appeal to her visitors' environmental identity, but her focus group participants suggest that, given the typical visitors' strong environmental identity, she could go further and include climate actions. Younger audiences may provide a window of opportunity for instilling positive climate change behaviors and attitudes, as their worldviews are not yet as entrenched as those of adults. Staunch climate change skeptics may be unlikely to change their opinions in part because climate change skepticism is linked to their identity.[35] Environmental educators may want to expend less energy trying to convince skeptics that climate change exists and more energy working with people who accept and are concerned about anthropogenic climate change but need assistance in deciding how to act on their concern. If educators work with climate change skeptics, they may want to take Elena's approach: find areas of common ground that enable both groups to work toward solutions even as they disagree on causes of the problem.

6

PSYCHOLOGICAL DISTANCE

Personal relevance is a keystone of environmental education, as well as a determinant of learner outcomes in climate change education programs.[1] One method of increasing relevance is to underline how climate change impacts local people and places—to make a global phenomenon a local phenomenon. Elena, Jayla, and Will employ place-based approaches, rooting their programs in local places and phenomena as part of their attempt to make climate change relevant to their audiences. Elena talks about climate change in terms of impacts to her town, while Jayla emphasizes climate change impacts to local native plant species. Will uses a Kentucky supplement to the *Southeastern Forests and Climate Change* module and works with students to develop a local, collective action plan so he can enhance the local relevance of his program.[2]

Research suggests that people in the United States tend to view climate change as a distant problem. They consider it to be temporally, spatially, and socially removed from the here and now, and to be uncertain at multiple levels.[3] People perceive climate change as something that will impact future generations and that threatens other regions far away.[4] Scientists are actively studying the effects of this "psychological distance" on climate change attitudes and policy support, and engagement or personal connection with the issue of climate change expressed as caring, motivation, willingness to act, and action itself.[5]

The idea of psychological distance lies at the core of construal-level theory.[6] Construal-level theory holds that psychological distance affects how people mentally represent, or construe, objects and events. People's construal, in turn, has important implications for how they use information in everyday judgment and decision making.[7] The theory posits that as events are perceived as psychologically closer (more proximal), they give rise to mental representations (construals) that are more concrete, more detailed, vivid, and contextualized. Likewise, as events are perceived as psychologically farther away (more distal), they give rise to mental representations that are more abstract.[8] For example, climate change impacts, such as drought or sea level rise, may be psychologically close for some but psychologically distant for others. Those living in towns especially vulnerable to these impacts may construe them in highly concrete ways—for instance, by imagining the feeling of parched grass crunching under their feet or lifting sandbags alongside their neighbors to construct makeshift levees. Those living far from these threats may think about them much less vividly.

One might assume that making climate change psychologically close, and therefore concrete, would enhance concern or the likelihood that someone would take climate action. Yet studies examining how distance affects people's engagement with climate change and their willingness to commit to action have yielded mixed results. A 2005 U.S. study found moderate levels of public concern, which was "driven primarily by the perception of danger to geographically and temporally distant people, places, and nonhuman nature."[9] Similarly, in a study conducted in the UK, researchers found that people were *more concerned* about the impacts of climate change on *distant* places. They also found, however, that messages about local impacts resulted in participants having more positive attitudes toward climate change mitigation.[10] These conflicting results might be explained by another factor, such as how attached one is to one's local place. A British Columbia study revealed that people who received messages about local climate change impacts and had higher levels of local place attachment reported higher levels of climate change engagement.[11]

Psychological distance also includes the distance that someone feels toward another person or toward society (social distance), a feeling that can interact with social and political identities. Emotions like compassion can play a role in decreasing this type of psychological distance.[12] In one study, using compassion appeals that asked people to imagine the experience of a child suffering from malnutrition because of drought increased conservatives' and moderates' compassion for others suffering from climate change impacts and, in turn, increased policy support for climate change action. Compassion appeals were less effective at increasing policy support among liberals, who already reported high levels of support.[13] Another study looked at how messages that portrayed victims of climate change impacts as living closer or farther away from study subjects affected support for mitigation policy, based on political party. While Democrats' support for mitigation policy *increased* when the victims were geographically distant from them, Republicans' support *decreased* when the victims were geographically distant and *increased* when the victims were geographically close but temporally distant. Results like these imply that effective messaging for conservatives might include highlighting future, local impacts.[14]

These mixed results of studies of psychological distance of climate change highlight the complex ways in which people process climate change information. They also suggest that although psychological distance research can inform education programs, simply including local impacts and other locally relevant information will not be a silver bullet to achieving outcomes like climate-friendly behavior.[15] Perhaps future research examining the interaction of distance with other factors will help to clarify these relationships.

Bottom Line for Educators

How to best apply research in psychological distance in environmental education is unclear, as the psychological distance research does not necessarily suggest that local is always best. For example, conservative audiences may be more likely to respond to messages that integrate compassion,[16] local area, and distant future.[17] Jayla combines a discussion of local and more regional climate change impacts, although her focus group wishes she would be even more specific about local impacts and action opportunities. Just as attitudes are important but not sufficient for motivating climate change action, personal relevance achieved through grounding climate change in local settings may be a piece of a larger climate change messaging puzzle. Drawing on people's sense of compassion for human victims of climate change impacts could increase concern for climate change in conservative audiences, whereas highlighting the local environment and local communities could help make climate change a more salient topic for audiences who have high levels of place attachment.

7

OTHER PSYCHOLOGICAL THEORIES

Psychological denial mechanisms, like terror management theory and cognitive dissonance, add to our understanding of the psychology of climate change communication. Together with identity (chapter 5) and psychological distance (chapter 6), these denial mechanisms round out our section on psychology.

Terror Management Theory

In the fall of 2017, author Anne Armstrong facilitated an online course called "Climate Change Science, Communication, and Action." Part of her work involved managing a Facebook group with fifteen hundred participants, many of whom posted alarming climate-related articles daily. By the end of the three weeks, Anne found herself so overwhelmed by the volume of apocalyptic climate change news that at one point she decided she might as well give up and just live life according to perceived U.S. norms—drive her car everywhere, forget about shortening her showers, and abandon feeling guilty about the energy she used washing her little girl's cloth diapers (and indeed, about having made the decision to have a child at all). In response to the "emotional labor" required to uphold a façade of hope and positive energy,[1] and to real fear about the future, Anne had put up an emotional defense system to manage her fear of climate change.

According to terror management theory, we spend our lives trying to survive and yet are faced with the persistent realization that we will eventually die. Because of this awareness of our inevitable mortality, when confronted with thoughts about death we engage in psychological defenses to ward off what could otherwise be a crippling "mortality salience." Climate change may provoke these psychological defense systems.[2] To counter the mortality salience that thinking about climate change evokes, people may focus on how unlikely it is that strong storm events or other climate change impacts will affect them and engage in "ego-protective processes," such as telling themselves it is highly improbable that such a storm would hit where they live. As news of deadly hurricanes, wildfires, tornadoes, and flooding, and their connection to climate change, becomes more common, such defensive processes may encourage irrational beliefs and behaviors that are, on their surface, seemingly unrelated to death. These beliefs and behaviors relate to reaffirming our sense of significance in the world. They can include bolstering our self-esteem by adhering more strongly to cultural symbols and group values.[3] For example, after reading about climate change threats, survey respondents from Austria reported that their intentions to engage in pro-environmental behavior decreased while their ethnocentrism increased.[4] Other self-esteem-bolstering behaviors in response to frightening climate messages could include the purchase of items like SUVs, which symbolize safety, stability, and success.[5]

Environmental education can provide alternative means to enhance self-esteem, such as stewardship activities that build self-efficacy.[6] Elena, Jayla, and Will try to moderate their audiences' fear responses by balancing descriptions of climate change threats with opportunities for action. In particular, Elena's and Will's stewardship projects provide opportunities for community building that help bolster people's self- and collective efficacy.

Cognitive Dissonance Theory

Cognitive dissonance theory suggests another means by which people might deny climate change or fail to engage in climate-friendly actions. According to this theory, individuals attempt to reduce negative feelings that accompany inconsistent (dissonant) attitudes and behaviors by changing either the behavior or the attitude, or by denying that any conflict exists.[7] Someone who holds pro-environmental values and flies frequently for work might feel an uncomfortable tension (called "dissonance") when thinking about her large carbon footprint, a tension she is motivated to reduce. To reduce this dissonance, she may commit to fly less and buy carbon offsets. Alternatively, she may relax her pro-environmental standards and even justify her behavior by denying that emissions are problematic or by telling herself that paying the bills and supporting her family are more important. As Per Espen Stoknes writes in his book *What*

We Think about When We Try Not to Think about Global Warming, "For my own part, I feel dissonance each time I fly. I still do it, though. It doesn't help much that I use my electric bike as much as I can when home. My own solution is to buy four times the amount of carbon quotas that I fly for, from the EU trading system. If I want to participate in our current society . . . I'll have to endure some inner dissonance."[8]

One way to combat cognitive dissonance is to provide audiences with actions they can take promptly and easily (see chapter 8). Stoknes purchases carbon quotas, but environmental educators have the capacity to involve people in actions directly through their programming. Each of the educators in the vignettes provides opportunities or examples of easily accessible mitigation and ecosystem-based adaptation actions. Elena runs a volunteer shoreline restoration program; Jayla's focus group suggests she link her exhibit to the center's rain barrel education program; and Will builds a tree-planting program into his climate change curriculum.

Bottom Line for Educators

Climate change programs risk activating people's terror management responses if they portray the issue as doom and gloom. Instead, educators should prioritize programs that inspire hope and help build participants' confidence in their capacity to be part of feasible climate solutions. Easy-to-implement actions that audiences can take on a daily basis, like biking, walking, or taking public transportation more often, may help reduce the cognitive dissonance that many feel when using energy and resources.

PART 2 RECAP

In this section, we examined four psychological processes that govern reactions to and engagement with climate change information. *Social identity* interacts with cognitive processes like motivated reasoning and identity protective cognition, leading people to draw conclusions that support their identity and to avoid those that threaten that identity. *Psychological distance* is an active topic of investigation in climate change communication and psychology research. However, because researchers have found mixed results, we suggest being on the lookout for new research, considering factors such as place attachment that may interact with messages emphasizing local and global distance, and paying close attention to what resonates with your particular audience. *Terror management theory* suggests that we deal with our awareness of our own demise by denying it or by seeking to bolster our self-esteem through adherence to cultural values and symbols, which may not help address climate change. *Cognitive dissonance theory* offers a perspective on the psychological gymnastics people do to deal with inconsistencies in their knowledge about climate change and their behavior.

Tips for Educators

1. Before developing a new climate change program, do the legwork to understand your potential audiences' attitudes toward climate change and the relevant social identities that may come into play when you discuss climate change. The National Audubon Society's *Tools of*

Engagement: A Toolkit for Engaging People in Conservation[1] and the North American Association for Environmental Education's *Guidelines for Excellence: Nonformal EE Programs*[2] are open-access resources that provide information on how to conduct audience assessments.

2. Whether messages about local impacts or about distant impacts will be more effective in achieving your climate change education outcomes varies. Choose message types that seem to resonate with your particular audience.
3. Use human stories to build compassion by asking audiences to put themselves in the shoes of people experiencing climate change impacts.
4. To avoid terror management responses, keep your messages hopeful and solutions-based.
5. Giving examples of easily adopted actions that audiences can participate in may help reduce cognitive dissonance.

Part 3
COMMUNICATION

Have you ever had a teacher who explained a fact to you in such an interesting way that you've never forgotten it? Perhaps you can still quote the exact words the teacher used to describe the phenomenon. Precise, well-crafted language that frames the message with the audience in mind, employs compelling metaphors, and comes from a trusted source helps audiences to understand concepts. Educators can use what they know about their audiences to inform their language choice and overall program organization, which in turn may facilitate achieving their outcomes. This part summarizes research on climate change communication and helps educators apply that research to their education programs. We focus on aspects of climate change messages, including framing (chapter 8) and metaphors (chapter 9), and on the messenger who delivers the message (chapter 10). We continue weaving in examples from Elena's, Jayla's, and Will's stories to illustrate applications to environmental education.

8

FRAMING CLIMATE CHANGE

"Framing" refers to how communicators use features of a message to evoke ideas and ways of thinking that audiences use to interpret that message.[1] Frames make different ideas more noticeable or important, or what communication researchers refer to as "salient." This in turn can affect how audiences assess information. In this chapter, we focus on emphasizing aspects of a message, including appeals that draw on an audience's preexisting knowledge, to guide their understanding. We can think of frames as "interpretive storylines that set a specific train of thought in motion, communicating why an issue might be a problem, who or what might be responsible for it, and what should be done about it."[2] As environmental educators, we frame the information that we share using a variety of strategies.

Communication researchers distinguish between two broad categories of frames: equivalency frames and emphasis frames. Both kinds of frames link concepts together, assisting the reader in recalling and interpreting ideas.[3] Emphasis frames use specific words to appeal to particular areas of an audience's knowledge or interest.[4] Take the statement "warmer temperatures are causing changes around the world, such as melting glaciers and stronger storms."[5] This statement frames climate change impacts as global, occurring "around the world," and as affecting natural features and systems like glaciers and weather. Let's contrast that with another statement about climate change and weather: "Do you live in the Northeast? You've experienced the very biggest rainstorms getting 70 percent bigger in the last 55 years."[6] This statement frames climate change impacts as local to the Northeast and as highly personal to the reader—perhaps as a means of making the message feel psychologically close. These different frames are likely to activate different thoughts or even different mental models people use to interpret the information.[7] Framing the statement about weather and climate in terms of the Northeast may activate an audience's mental representations of what it's like to live in the Northeast, whereas the mention of big rainstorms can evoke stored knowledge and memories of previous experiences with major precipitation events.

Like emphasis frames, equivalency frames draw attention to certain aspects of a story. Equivalency frames present logically equivalent information but emphasize one part of the information to affect preferences.[8] A familiar example comes from the meat counter at the grocery store, where meat packages labeled as "80 percent lean" outnumber (and presumably outsell) meat labeled "20 percent fat." Because people evaluate statements based on what is emphasized, they are inclined to view the same information in a more positive light when it deemphasizes fat content.[9] In the context of climate change, the statement "97 percent of scientists agree that global climate change is happening" could be equivalently phrased as "3 percent of scientists do not agree that global climate change is happening." While the statements may be equivalent in a logical or mathematical sense, they emphasize different things—the first statement emphasizes consensus, whereas the second emphasizes contention—which carries implications for how audiences will process the message.[10] Consensus frames have been shown to increase people's acceptance of anthropogenic climate change,[11] and the "97 percent" frame better underlines scientific consensus on climate change.

Framing is ubiquitous in everyday communication and an active area of research across the social sciences, including in communication, political science, psychology, and behavioral economics. A good source for framing information is the National Network for Ocean and Climate Change Interpretation (NNOCCI), which draws on this research and research conducted by the FrameWorks Institute to provide guidance for climate change educators on how to frame their programs. Below we cover research on frames used in climate change communication, including framing around identities, and for hope, self-efficacy, solutions, values, and particular audiences. We also discuss how to apply this research to environmental education.

Frames Used to Communicate about Climate Change

Most adults learn about climate change and other scientific issues from the media. Thus, examining media frames can help shed light on whether and how the public chooses to act to address climate change.[12] The media (and many environmental groups) use predominantly negative, doomsday scenarios when framing climate change (table 8.1). They also provide few practical and effective actions for the audience to take, which may lead audiences to tune out the message—a problematic outcome for communicators and educators.[13] Frames may emphasize the economic risks or benefits of climate change or present climate change as a moral or ethical issue. Other frames emphasize scientific uncertainty or underscore the scientific consensus surrounding climate change impacts. Doomsday scenarios might help to gain the public's attention, but without clear solutions that audiences can implement, appeals to fear often fail to inspire action. In an attempt to adhere to the journalistic norm of balanced coverage, the media have also been criticized for giving equivalent voice to climate skeptics and framing climate change in terms of debate, controversy, and uncertainty, thus adding to the public's perception of a lack of scientific consensus.[14]

TABLE 8.1 Common climate change frames and examples from the media, adapted from Matthew Nisbet, "Communicating Climate Change"

FRAMES	EXAMPLES FROM MEDIA HEADLINES
Economic development and competitiveness	"Climate Change Will Be an Economic Disaster for Rich and Poor, New Study Says"[15] "Gambling the World Economy on Climate: The Emission-Cut Pledges Will Cost $1 Trillion a Year and Avert Warming of Less Than One Degree by 2100"[16]
Scientific and technical uncertainty	"Climate Science Is Not Settled"[17]
Doomsday, tipping point	"Climate Catastrophe Will Hit Tropics around 2020, Rest of World around 2047, Study Says"[18]
Morality and ethics	"Is the Environment a Moral Cause?"[19]

Studies of framing in climate change education provide insight into students' reactions to frames and how educators use frames (table 8.2). A German study found that undergraduate students who read sensationally framed information about climate change exhibited higher levels of knowledge retention than did students who read neutral information.[20] The sensational frame also increased student perception of climate change risk, and in turn led to stronger negative emotions toward climate change. However, terror management theory (see chapter 7) would predict that the negative emotions elicited by the sensational frames might hinder students' willingness to act on climate change. A study of how U.S. environmental educators use climate change communication research in their practice found that most educators use local frames, focus on solutions as a way of inspiring hope in their audiences, and view science frames as aids in maintaining political neutrality.[21] Finally, a study in California found that science teachers predominantly use scientific and global frames to discuss climate change (see table 8.2).[22] Although the researchers did not test the effectiveness of different frames, they suggested based on previous literature that including the human side of the issue may be more effective than science frames at activating the emotions that drive action.[23]

TABLE 8.2 Climate change frames used in environmental education programs, adapted from K. C. Busch, "Polar Bears or People?" and Anne Armstrong, "Climate Change Communication in Environmental Education"

FRAME CATEGORIES	EXAMPLES
Global impacts	Warmer temperatures are causing other changes around the world, such as melting glaciers and stronger storms.[24]
Local/proximal	Addressing climate change in ways that the audience has actually seen—more insect outbreaks, devastating wildfires, and how forests are being managed.[25]
Science based	Average global sea temperature has been rising gradually over several decades, 0.7°C in the past thirty years alone, which is generally believed among the scientific community to be due to global warming.[26]
Human-impacts based	Climate-related changes can make it difficult for Inuit hunters to reach the places where they hunt.[27]
Collective solutions	Pointing out things that are going on right now that people are doing; for instance, in our county, many businesses and organizations have actually have been very successful promoting renewable energy.[28]
Individual solutions	Deforestation to plant palm plantations adds to climate change, so make sure to buy products labeled with "Roundtable of Sustainable Palm Oil (RSPO)" and help promote sustainable palm production.[29]

Framing with Audience Identities in Mind

Although subtle differences in wording may seem trivial, they can have important effects on how people process climate change information. For instance, research suggests that the public responds differently to the labels "global warming" and "climate change."[30] Survey experiments conducted in the United States find that the public reports greater belief in the existence of "climate change" as compared to "global warming," an effect that is especially pronounced among Republicans and conservatives.[31] U.S. conservatives also have been found to associate more heat-related climate impacts with "global warming" than with "climate change," whereas liberals associate impacts with both phrases equally.[32] Another study further demonstrated that the use of the term "global warming" reduced Republicans' but not Democrats' belief in climate change and weakened both groups' perception of scientific consensus.[33] Not only does choosing one phrase over the other affect an audience's interpretation based on their social identity; the phrases themselves are not scientifically interchangeable. Global warming is the rise in average global-level land, air, and water temperature, and climate change results from these warming temperatures.

Religion and morality have recently received attention as possible focal points of climate change messaging. Although some research suggests that climate change fails to activate our moral judgment,[34] following Pope Francis's release of his climate change encyclical, *Laudato Si*, Americans were more likely to consider climate change a moral or ethical issue.[35] Additional research suggests that even brief exposure to a picture of the pope and a statement about his views on climate change increased the likelihood of regarding climate change as a moral issue among Democrats and Republicans.[36] This finding implies the pope may be a figure who bridges political boundaries, although in another study, conservatives who were aware of the papal encyclical were less likely than liberals to view the Pope as a credible source of climate change information.[37] Religious frames may offer environmental educators new modes of communicating about climate change and may open opportunities for organizational partnerships.[38] Educators interested in developing programs around faith may want to connect with local faith leaders who are trusted sources of information and values.

Environmental educators appeal to their audiences' regional, professional, and interest group identities through framing.[39] Educators who trained foresters using Project Learning Tree's *Southeastern Forests and Climate Change* framed climate change around impacts to forest health in their region, whereas educators working with municipal leaders appealed to climate change action as preserving quality of life in the community. An educator working with bird-watchers framed climate change in terms of the impacts on the birds that her audiences love. Another educator appealed to his audience's religious identities by creating a skit that framed climate change action around the idea that God wanted humans to be good stewards of the earth. Recognizing social identities is an important part of "knowing your audience" and an important strategy for developing suitable program language and content.

Framing for Solutions

Framing climate change solutions is as important as framing information about climate science and impacts. When the media do present solutions, the solutions rarely match the scale of climate change as a global and intergenerational issue. The problem with this approach is "that it easily lapses into 'wallpaper'—the domestic, the routine, the boring and the too-easily ignorable. It can be lacking in energy and may not feel compelling. It is often placed alongside alarmism—typified by headlines like '20 things you can do to save the planet from destruction.'"[40] Solutions that do not appear to match the significance of the threat can deplete people's sense of "response efficacy," or their perception of whether recommended actions will address the problem. As climate change presents threats at different scales, educators who do focus on the global scale would want to offer suggestions for how to link to global action. Similarly, educators highlighting local climate change impacts can focus on local climate change actions.

Although individual behaviors are certainly important and have the potential to substantially reduce greenhouse gas emissions,[41] collective actions may feel more appropriate given the scale of the problem. One way to bridge individual and collective actions is for audiences to share their individual actions through social networks—helping to make that action part of a collective movement.[42] Returning to our vignettes, Jayla takes this approach and recommends that visitors support local renewable energy initiatives in their city—an individual action that, if performed by many, will promote renewable over fossil fuel energy at a regional level. Will and Elena both plan collective stewardship actions, while Jayla includes suggestions for collective action in her exhibit. Will's students plan a tree-planting program, and Elena recruits a group of volunteers to create a living shoreline.

Citizen scientists in one study exhibited increased interest in a carbon footprint activity when they read messages that were framed in terms of collective action.[43] As an example of this in practice, NNOCCI trains educators to highlight community-level solutions in their programs, like joining community renewable energy collectives or working within their communities to enhance bike transportation. Research suggests that successful campaigns for individual or household-level climate action also include a social dimension in the form of marketing through social networks—making the individual who takes action feel like part of a collective movement.[44]

Framing for Self-Efficacy and Hope

NNOCCI educators cited community-level solutions as a means of providing their audiences with hope.[45] Feelings of hope, along with self-efficacy, are related to willingness to engage with climate change information.[46]

Self-efficacy is a "foundation for [environmental] action" because it "contributes to a sense of self-worth and resolve necessary to set and reach challenging goals."[47] Communication researchers have tied self-efficacy to climate change action. People feel a greater sense of self-efficacy and believe their actions will decrease climate change impacts if they receive messages that frame climate change in terms of what they gain from action versus what they lose from inaction (e.g., "If *we act, we can* mitigate climate change impacts," versus "if we *don't* act, we *won't be able* to mitigate change impacts."). In short, more positive statements may better promote self-efficacy.[48]

Whereas self-efficacy is the expectancy that you can meet a goal, hope involves not only the expectancy but the ways in which to achieve the goal.[49] Hope consists of goals (what we want to happen), pathway thinking (our ability to figure out how to meet those goals), and agency thinking (motivation to use those pathways).[50] Sources of hope for Swedish and U.S. high school students include trust in themselves as individuals, trust in others, and positive reappraisal (cognitively reframing something as more positive).[51] High school students in North Carolina were more likely to engage in pro-environmental behavior if they were hopeful about climate change solutions.[52] For Swedish teenagers and young adults, hope in one's own and others' ability to meet the challenges of climate change predicted higher rates of energy conservation. Other emotions like worry may interact with hope to promote behavior. For example, among Swedish adults, hope predicted environmental behaviors like recycling, but only for people who were worried about climate change.[53]

How does one frame for hope? In addition to focusing on positive, solutions-based messaging, educators may want to consider framing climate change in terms of public health. Communication researchers tested the effects of framing climate change in terms of public health, national security, and risks to the environment on emotions including hope and anger. Authors connected climate change actions, like redesigning cities for safer foot and bike travel and public transportation, to reductions in traffic injuries. They also linked these actions to the benefits of increased physical activity and to the accompanying reductions in carbon emissions. The health frames evoked the most hope among political independents and conservatives. Moreover, conservatives and independents were more likely to support climate change mitigation when they read about climate change in terms of public health.[54] These findings suggest that some segments of the public may find health a more approachable, tangible subject than climate science, thus eliciting more hopeful feelings and inspiring climate action.[55] Framing environmental concerns around public health may be familiar to some environmental educators. For example, Project WET incorporates public health into watershed lessons, such as learning about hand-washing or solving a mystery about the origins of a cholera epidemic in nineteenth-century London.[56]

Framing for Values

Researchers have explored which values influence climate-friendly behavior, and environmental groups have started promoting values-based messaging as a way of targeting particular audiences.[57] Values play a role in defining social identities, such as political or religious identities, and can be another piece of the puzzle that explains climate change attitudes.[58] Values serve as guiding principles in our lives; they can be acquired through interactions with social groups and through individual experiences.[59] Some examples of values that we hear about in day-to-day conversation include family values or environmental values. Three types of values, altruistic (focus on the welfare of other people), biospheric (focus on the welfare of the environment), and egoistic (focus on oneself), have been used to help explain pro-environmental behaviors.[60] You might join a community renewable-energy collective for a variety of reasons that correspond to any one of these values; for example, you might perceive joining the collective to be the most affordable option (egoistic), believe that renewables help mitigate climate change impacts on human communities (altruistic), or you might see climate change mitigation as a way of minimizing impacts on the environment (biospheric).

Early in the climate change movement, groups tailored their messages to the pro-environmental values that their members held, focusing on "save the earth" messages or biospheric values.[61] Intuitively, you might expect biospheric or altruistic values to be especially predictive of environmental behavior, but research findings are mixed. A study conducted in the UK identified altruistic values (in particular, values related to social justice) as stronger motivators of low-carbon behavior than biospheric values.[62] Research involving residents from Michigan and Virginia found that altruistic values, along with traditional and family values like honoring one's parents or showing respect, correlated with support for climate change mitigation policies.[63] Some researchers argue that seemingly altruistic acts may actually be motivated by self-interest alongside altruism.[64] As opposed to selfishness, or operating without concern for others, self-interest is about taking care of your needs so you can continue to function and achieve a sense of happiness. Happiness can be tied to a variety of outcomes, including seeing others succeed and environmental improvement. When Elena says, "We all want to protect this area," she appeals to town members' altruism and self-interest.

Framing that appeals to strongly held values can provide a shortcut for audiences as they judge how and whether information is relevant to them.[65] A study of forest landowners tested four different videos framed around stewardship (a biospheric value) and timber (an egoistic value). When viewers' values aligned with the particular frames in the videos, they reported liking the video, trusting the messenger in the video, and they registered higher rates of intending to take forest stewardship actions and actions to address climate change.[66] Other research suggests that framing pro-environmental appeals to resonate with national-level cultural values can bolster environmental engagement, as when appeals to purity increased purchases of carbon offsets among airline travelers from India, whereas appeals to individual choice increased such purchases among Americans.[67]

Beyond Word Choice—Developing a Program around Frames

You may wonder how to frame climate change in an entire program, beyond strategically choosing a few words or phrases. Fortunately, frames do more than elicit a particular interpretation; they can also communicate about a problem, its causes, and solutions. This conception of framing links to the work of the sociologists Robert Benford and David Snow on framing for social movements. Their explanation of core framing tasks is helpful for considering how to frame messages throughout an environmental education program.[68]

The first core framing task is to identify the problem and to explain who or what caused the problem. This is called "diagnostic framing." In a climate change education program, this could take the form of the introduction to a main climate-related topic covered in the program (e.g., climate change impacts on a local endangered species or on human health) and why climate change is occurring. The second core task is to propose solutions to the problem, or "prognostic framing." In a climate change education program, this might include a discussion of what your organization or people in the community are proposing to do about climate change and even a discussion of proposed solutions that you think will be ineffective. While prognostic framing proposes solutions, motivational framing (the third core task) is a call to action that encourages audiences to be agents of change who work toward those solutions. These core framing tasks could take many forms in an educational program, from lectures to dialogue to games to actual climate change mitigation or adaptation actions

Elena demonstrates how an educator can focus on intended outcomes (her town taking collective action to adapt to climate change by restoring coastal ecosystems) and can adapt her message depending on the audience (figure 8.1). She starts by involving her audiences and asks them to define the problem by sharing their experiences locally with coastal flooding and land change (diagnostic framing). She then explains why the problem exists. She uses a "climate change" rather than a "global warming frame" and emphasizes impacts related to sea level rise. Although she meets with resistance to her "diagnosis" about what

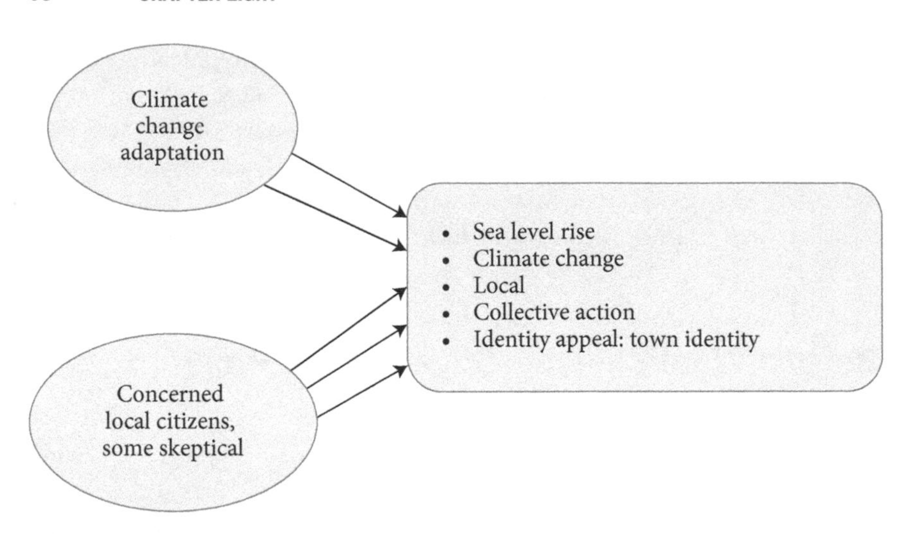

FIGURE 8.1 How Elena's outcome and audience inform her frames

causes flooding, she is still able to motivate her audiences by focusing on how they can address flooding through volunteering at a living shoreline event. In addition, Elena elicits her audience's local "town" identity; protecting the shoreline becomes a way to protect the town and its identity. Finally, Elena focuses on a collective action rather than individual behavior to inspire collective efficacy and the expectation that the town will meet climate change adaptation challenges in an environmentally sound way.

Jayla's outcome is climate change literacy, and she chooses a mix of regional and local frames to describe climate change (for example, she writes about climate change in California and in her nature center). Her nature center audience is concerned about climate change, but her volunteers explain that she needs to make a stronger case for why visitors should care about chaparral by appealing to their biospheric values and highlighting the uniqueness of chaparral systems.

Will seeks to build student climate literacy but also wants to achieve positive youth development through his students' action projects. Knowing some of his students might resist discussing climate change because of their conservative social identity, he begins by asking students to consider each other's opinions, hoping this will make all students comfortable discussing their ideas about climate change.

Bottom Line for Educators

Just as educators choose particular activities to meet their educational outcomes, they can choose frames that aid in achieving those outcomes. Choosing between "coastal flooding" and "climate change" has implications for how particular audiences interpret the message, and therefore their overall understanding and willingness to take action. Building messages around hope and self-efficacy also enhances the likelihood audiences will act. Framing using values can provide a shortcut for audiences as they judge whether information matters to them. Educators can also organize their programs around diagnostic, prognostic, and motivational framing tasks. In general, using a positive frame by showing how people can take collective action to address a local problem is good practice.

9
USING METAPHOR AND ANALOGY IN CLIMATE CHANGE COMMUNICATION

Similarly to frames, metaphors and analogies facilitate communication and comprehension. Metaphors ground abstract concepts in concrete physical reality and involve an implicit comparison between concepts that are unrelated but share some common characteristics.[1] "Raining cats and dogs" and "the elephant in the room" are examples of common metaphors. These metaphors activate mental representations that structure how people perceive the message.[2] We know that it will not rain cats and dogs in a literal sense, but the metaphor conveys the essential point that it is raining very hard. We know that there is not actually an elephant in the room, but rather that there is a large looming problem people would rather not discuss.

Educators use metaphors not only to convey meaning but also to enhance student retention of information by connecting abstract concepts to known concepts.[3] Climate change communication is replete with metaphors. There are "hothouses and greenhouses, atmospheric blankets and holes, sinks and drains, flipped and flickering switches, conveyor belts and bathtub effects, tipping points and time bombs, ornery and angry beasts, rolled dice, [and] sleeping drunks."[4] The FrameWorks Institute tested a series of metaphors and found that the metaphors "rampant versus regular CO_2," "osteoporosis of the sea," and "climate's heart" were most successful at building an understanding of climate change.[5] NNOCCI adopted these metaphors as well as "heat-trapping blanket" in their training program. (See table 9.1 for examples of climate change metaphors and analogies.)

While they are often effective, metaphors can be tricky to use because they highlight similar but ignore dissimilar features between two subjects.[6] Take the greenhouse gas example. People understand the role of greenhouse gases in the atmosphere as acting like the panes of glass in a greenhouse: they trap heat. At the same time, a greenhouse has some characteristics that are very different from the earth's atmosphere, and the persistence of the metaphor may lead to misconceptions about the mechanics of global warming and the time scale on which we can stop warming trends. For example, to let heat out of a greenhouse, you simply open a door. Even if we stopped emitting CO_2 today, it would take millennia for the atmosphere to return to preindustrial levels of CO_2.[7] This metaphor also overlooks the important role of other carbon sources like methane and sinks like the ocean.[8] Elena and Jayla both chose the metaphor of "heat-trapping blanket" to describe how certain gases trap heat in the atmosphere. Of course, the "blanket" metaphor breaks down as well, as you can simply remove a blanket, but you cannot so simply remove gases from the atmosphere.

Analogies are similar to metaphors in that they draw comparisons between two ideas or objects. Analogies compare similar features of two domains. Will tries the analogy of atmosphere-as-bathtub into which humans continue to pour CO_2 at a faster rate than nature can drain the tub (figure 9.1). Some researchers distinguish between analogies and metaphors by explaining that analogies provide more explicit mapping of the similarities between A and B, whereas metaphors make the same comparisons implicitly.[9] Others explain that whereas metaphors claim that "A *is* B," analogies point out how "A is like B."[10] In practice, the distinction between an analogy and a metaphor is blurry. For example, an educator could say, "The climate system is a big carbon bathtub," and the statement could be classified as a metaphor. However, the educator could just as easily have used an analogy—"the climate system is like a carbon bathtub." The important point for educators is not to try to distinguish between the two, but to use the comparisons they make to help audiences understand climate change concepts.

Metaphors and analogies contain what is called an analog concept and a target concept. In the metaphors above, "cats and dogs" and "elephant" are the analog concepts, while raining especially hard and looming problem are the target concepts.

A metaphor or analogy on its own does not provide sufficient information for learners to build an accurate understanding of a concept. Imagine if you were giving a climate change education lesson and you said, "The climate system works kind of like a bathtub" and then just stopped there. Your audience would have no clear understanding of your meaning.

TABLE 9.1 Examples of climate change metaphors and analogies

METAPHOR OR ANALOGY	EXPLANATION
Carbon bathtub[11]	The bathtub represents the climate system, and the water level represents CO_2. Adding water from the tap represents addition of greenhouse gases into the atmosphere from human sources. If you keep adding water, eventually the tub will overflow unless you pull the drain to let some water out. More water will need to leave the tub from the drain than is entering the tub from the faucet to reduce the total amount of water in the tub, unless we can "drain" some of it out through carbon sinks. Similarly, the atmosphere will "overflow" with CO_2 unless we "drain" enough of it through carbon sinks.
Greenhouse effect	Gases in the atmosphere such as CO_2, nitrous oxide, and methane trap heat radiated from the earth just as the panes of a greenhouse trap the sun's heat.
The atmosphere as the climate changes akin to a baseball player on steroids[12]	Just as steroids increase the likelihood of hitting a home run in baseball, climate change increases the likelihood of severe weather. However, it is difficult to attribute a single home run to steroid use, just as it is difficult to attribute a single storm to climate change.

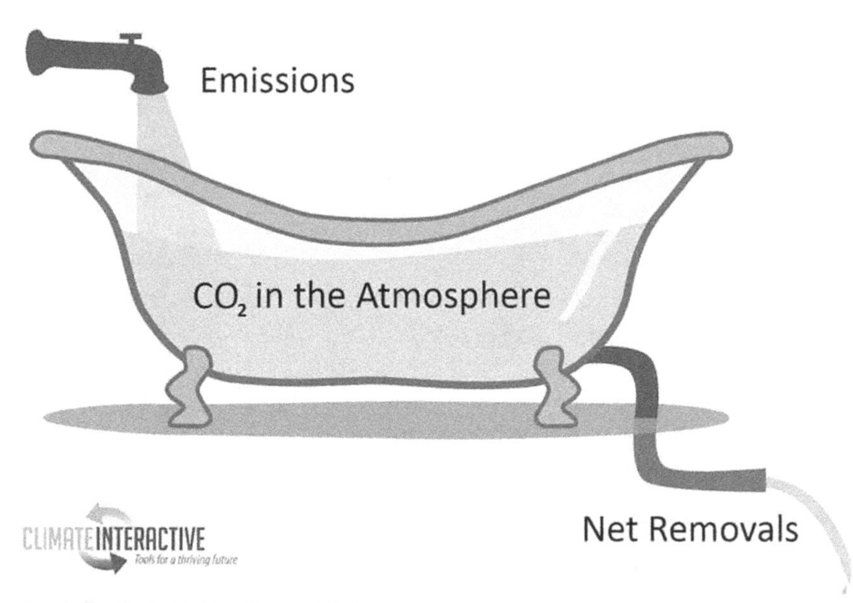

FIGURE 9.1 Carbon bathtub analogy
Climate Interactive[13]

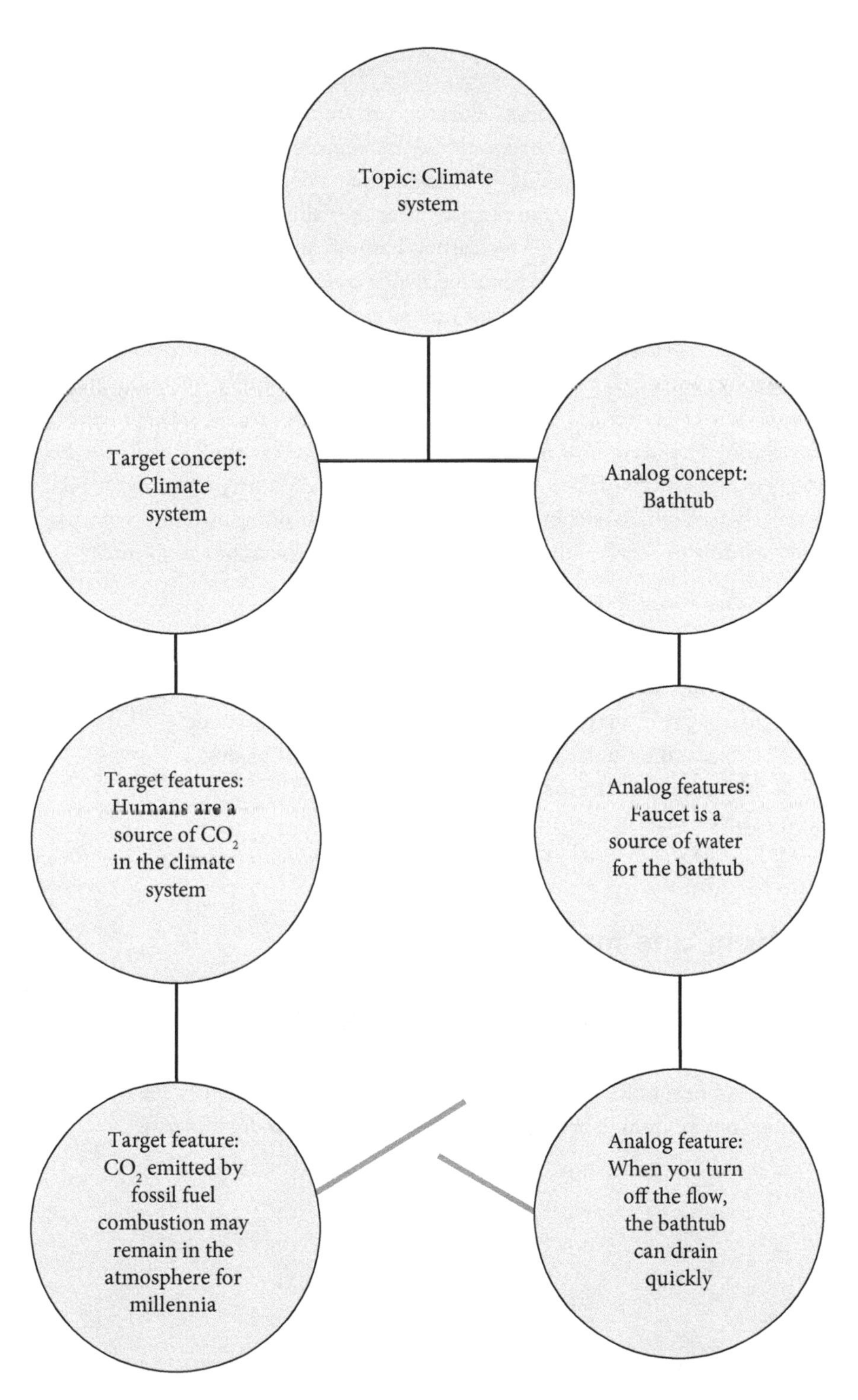

FIGURE 9.2 Elaboration of the carbon bathtub analogy. Broken lines represent the breakdown of the analogy.

Adapted from Shawn Glynn, "Making Science Concepts Meaningful to Students"

To strengthen comprehension, educators can embed analogies or metaphors in what researchers refer to as "elaborations."[14] An analogy elaboration "maps verbal and visual features of an analogy's analog concepts onto those of the target concepts," thus helping students to connect what they already know with new information.[15] In an elaboration of the carbon bathtub analogy, the target concept is how the climate system works, while the analog concept is the bathtub (figure 9.2). Educators who use this analogy want students to understand the idea of sinks and sources. In an elaboration, educators not only highlight the features that connect the analog concept with the target concept; they also identify where the analogy breaks down (figure 9.2).[16] In the bathtub model, this process aids audiences in developing a more accurate mental model of the target concept.[17] Visuals also help learners map aspects of the analog concept to the target concept (e.g., figure 9.1).[18] Similarly, the FrameWorks Institute recommends embedding metaphors in what it calls "explanatory chains" that present the causes and effects of a problem.[19]

Steps for elaborating a metaphor or analogy[20]

1. Introduce target concept
2. Introduce analog concept
3. Identify relevant features of both the target and the analog
4. Connect the similar features of the target and the analog
5. Indicate where the metaphor breaks down
6. Draw conclusions

Bottom Line for Educators

Metaphors and analogies help audiences connect what they know already with a new concept. Educators should think carefully about how to elaborate metaphors and analogies to facilitate understanding of climate concepts and avoid perpetuating misconceptions. One way to do this is to carefully identify the target and analog concepts and then map out their similarities and dissimilarities.

10

CLIMATE CHANGE MESSENGERS

Establishing Trust

A beautifully crafted message coming from the wrong messenger can misfire because people doubt information from sources they distrust.[1] What does it mean to trust someone? In climate change education, we can think of trust as a psychological state where the trustor or student (let's say his name is Joe) accepts some form of vulnerability, and a trustee or educator (let's say his name is Lucas) communicates information.[2] Joe relies on, or trusts, Lucas to communicate accurate information about climate change. Joe is doubly vulnerable in this situation; if his expectations about Lucas's knowledge and capabilities aren't met, Joe will be disappointed and might also feel personally betrayed.[3] Trust has received growing attention as a central issue in promoting public understanding and acceptance of climate science.[4]

People tend to trust messengers who hold views similar to their own.[5] Such messengers can pass on climate change messages that "speak directly" to their peers by serving as "connective communication tissue, apprising peers about what is important."[6] For example, U.S. forest landowners responded more positively to climate-related videos that portrayed people who espoused their values.[7] Cooperative Extension educators also can serve as trusted sources of climate change information, especially when their backgrounds are similar to those of their audiences.[8] This suggests that in order to gain trust, environmental educators should partner with opinion leaders or communicators with backgrounds and values similar to those of their audiences.

In an analysis of zoo visitors' climate change attitudes using the "Six Americas" audience segmentation,[9] researchers found that even those visitors who were "cautious" and "disengaged" moderately trusted zoo educators as climate change information resources, while those in the "concerned" and "alarmed" categories demonstrated a high level of trust in zoos as climate change information sources. Nevertheless, sometimes seemingly small details can affect the credibility of a messenger. Zoo visitors in another study rated zoo employees with the titles "field biologist" and "animal researcher" as the most credible sources of climate and other conservation messages.[10] They rated "volunteers" and zoo "administrators" as having low credibility. Considering the fact that volunteer docents conduct much of the education at zoos and similar institutions, the study's authors suggest institutions use a term like "zoo educator" to convey volunteers' training and expertise.

Environmental educators viewed positively by the local community are in a good position to connect their audiences to climate change information and action, but when this is not the case, they can bring in a trusted member of the community. Elena holds her meeting in a community space rather than at her education center and hopes that the fire hall's approval of her event will translate to audience members extending their trust to her.

Bottom Line for Educators

Trust between the educator and the audience plays a key role in audience receptivity to climate change messages. Educators can establish trust by working with local trusted partners and opinion leaders and by thinking carefully about the messenger. Luckily for environmental educators, their audiences may already consider them as trusted information sources.

PART 3 RECAP

Research on frames helps us develop and choose language that resonates with audiences and motivates action and support for policy. Positive frames that highlight collective action may promote hope and self-efficacy, which are precursors to action. Metaphors and analogies provide additional tools that aid educators in communicating abstract concepts to their audiences. In addition to the message, the messenger is important, because trust plays a role in determining audience reactions to climate change messages. Crafting a message strategy, which includes what the message is and who the messenger is, is integral to program planning and achieving intended outcomes. To craft strategies effectively, educators should apply what they know about their audiences to create messages and programs that appeal to their particular audience's identities and values.

Tips for Educators

1. Before you begin developing frames, think about your audiences' values and knowledge. What will resonate most and seem most relevant to your audience?
2. Test out new ideas in program planning. Frames will resonate with your audience in different ways. Prior to starting your program, test out a few framing approaches to see which one your audiences respond to. Reflect on your results and refine your program.

3. Frame climate change in a hopeful and empowering way to avoid engaging terror management responses. One way to achieve this is by framing climate change through collective action.
4. Use metaphors and analogies to create connections between your audience's understandings of concrete issues with their understanding of climate.
5. Consider partnering with an opinion leader or trusted messenger who can help you establish credibility with your audience.

Part 4

STORIES FROM THE FIELD

Anne interviewed four educators and community leaders who served as Community Climate Change Fellows to learn about their teaching and communication strategies. The Community Climate Change Fellowship Program was part of Cornell University and the North American Association for Environmental Education's EECapacity environmental education training program, funded by the U.S. Environmental Protection Agency. The fellows' profiles represent approaches to climate change education across different settings and reflect the research presented in the previous chapters. At the end of each profile, we include tables to summarize the educators' approaches to climate change and connect their strategies with environmental psychology and communication concepts.

11

CLIMATE CHANGE EDUCATION AT THE MARINE MAMMAL CENTER, SAUSALITO, CALIFORNIA

Zoo and aquarium visitors are more concerned about climate change than is the general U.S. public, and those visitors who are most concerned are also readiest to participate in consumer or home-based behaviors to mitigate climate change.[1] Carefully designed exhibits and interactions with visitors can harness visitors' emotional connection with the zoo or aquarium animals to make climate change a personally relevant issue.[2] Here we tell the story of Adam Ratner bringing climate change interpretation to the Marine Mammal Center in Sausalito, California.

Weaving climate change into his work as guest experience manager at the Marine Mammal Center seemed like a natural fit for Adam Ratner. Adam has a background in biology and psychology and studied bird behavior before coming to the Marine Mammal Center. The center is the world's largest marine mammal hospital and education facility and is open to the public, reaching more than one hundred thousand people a year. Through his Community Climate Change Fellows project, Adam sought to "figure out ways that we can use climate change science and communication tactics to inspire change in all of our guests and communities." He created a volunteer training program modeled after the National Network for Ocean and Climate Change Interpretation (NNOCCI) program he had participated in. Adam's goal was to empower volunteers with "the skills, the comfort level to really share the stories of our animals and raise awareness and inspire people to take action." To accomplish his goal, Adam trained 125 Marine Mammal Center education volunteers.

In framing climate change, Adam emphasizes the values of protection and responsible management, which are relevant to marine mammal rehabilitation and which NNOCCI has found effective for engaging zoo and aquarium visitors with climate change. According to Adam, "At the Marine Mammal Center, protection is something that is relatively straightforward with the work that we're doing. The message is that we can take action today to protect and preserve important habitats for marine mammals and people alike. And then we can go into what is affecting those habitats, and what can we do to actually protect these animals."

What does Adam want audiences to take away from their visit to the Marine Mammal Center? "We're looking for systematic or community-level solutions for climate change in particular, so that the actions we take are at the same scale as what the problem is rather than something like turning off lightbulbs, which might not resonate very well for guests when we talk about what the scope of climate change impacts are. Turning off lightbulbs just doesn't really seem to meet that need."

This is where Adam makes use of the Marine Mammal Center exhibits and facility. The center has plenty of "visual examples that we can [use to] highlight [the steps] that we're taking to reduce our carbon footprint." The center has installed solar panels over the animal pens, which generate electricity, provide shade for the animals, and save money (solar is cheaper than fossil fuels in California). Adam trains his volunteers to use the solar panels as one example of a renewable energy solution audiences can "then take back to their own houses, their own businesses and see the benefit."

Adam explained that one of the keys to effective climate change interpretation at his site is "giving people a sense of hope. As we talk about what's making these animals sick it can get to be very overwhelming at times. We always try to make sure that when we talk about the issues, whether it's climate change or overfishing or ocean trash, that we're always building in what's next, what can we do to help this so that people feel empowered, inspired, and that they can actually prevent these animals from getting sick in the future."

When he was asked to give an example of how he or one of his volunteers would connect climate change with elephant seal survival, his response incorporated metaphors, elaboration, solutions, and connecting to audiences' values (figure 11.1).

Adam has, of course, faced some challenges in implementing his climate change education programs. One of Adam's biggest challenges has been developing a realistic sense of what trainees could commit to in terms of weekly homework assignments applying program concepts, and his own limits as a facilitator in trying to keep trainees on top of assignments. He realized that his time was best spent supporting "the people that are actively engaged in learning and practicing

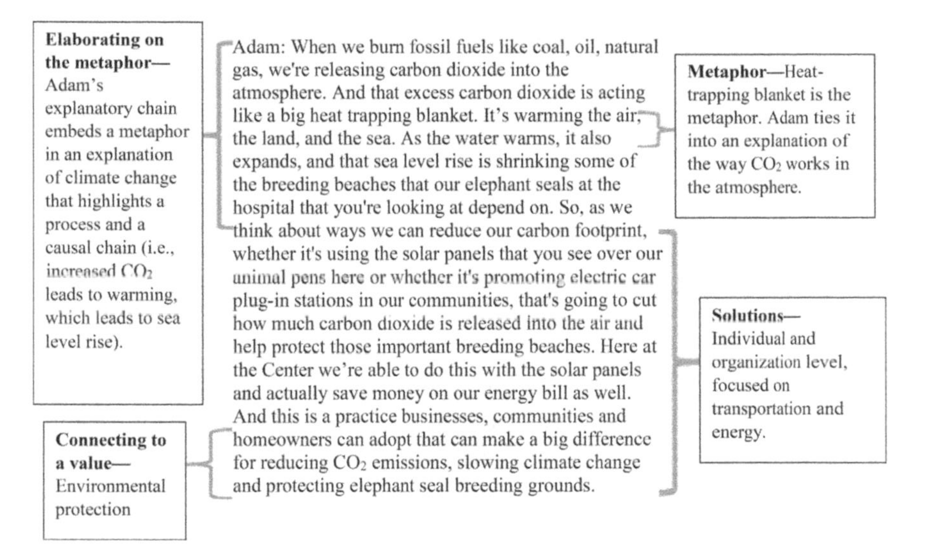

FIGURE 11.1 An example of how Adam Ratner interprets climate change for visitors at the Marine Mammal Center

new skills" rather than worrying about those who were not as committed to the program. As Adam's climate change education efforts have expanded beyond the Marine Mammal Center to developing regional workshops with the Bay Area Climate Literacy Impact Collaborative (BayCLIC), he has learned that "setting up a massive collaborative takes patience. We spent close to a year on logistics like a mission/vision statement, organizational structure [working groups, a steering committee] before we were able to really get going on some of the action that we were so motivated and excited to get to from the beginning." Adam and his collaborators' up-front efforts have given them confidence in BayCLIC's direction moving forward.

Summary

Adam frames climate change around the "heat-trapping blanket" metaphor, his audiences' values (protection and responsible management), and community-level solutions. He uses these communication strategies to instill hope in his volunteers and the audiences so that they go home and pursue climate-friendly actions (table 11.1).

TABLE 11.1 Summary of Adam Ratner's climate change programs and how they connect to concepts covered in chapters 1–10

SETTING	THE MARINE MAMMAL CENTER, SAUSALITO, CALIFORNIA
Audience	Direct: Volunteers Indirect: General public
Primary program outcomes	Climate literacy Collective action
Climate psychology connections	Appeals to audience values Uses psychologically close examples of solutions Appeals to audience emotions like compassion for animals
Climate communication connections	Uses positive, hopeful messaging Uses metaphors like "heat-trapping blanket" Embeds metaphor in an explanatory chain
Program strategies	Trains volunteers to deliver climate change interpretation at exhibits following NNOCCI guidelines

Adam's Tip for Educators

Give audiences a sense of hope by building in a discussion of community-level, collective actions they can take.

12

CLIMATE CHANGE LITERACY, ACTION, AND POSITIVE YOUTH DEVELOPMENT IN KENTUCKY

Like environmental education, climate change education can have multiple outcomes, from climate action to positive youth development. Environmental action involves decisions, planning, implementation, and reflection by an individual or group to achieve a specific environmental outcome,[1] whereas climate action refers to an environmental action with a specific climate adaptation or mitigation goal. Similar to environmental action, climate action provides opportunities for positive youth development.[2] Below we tell the story of Jennifer Hubbard-Sanchez's work at Kentucky State University, which involves undergraduate students in climate action projects that lead not only to climate action but also to positive youth development outcomes like leadership, taking responsibility, and building connections with others.

How would you approach climate change if you lived in a state whose leading politician says that "each side has their scientists"[3] and where the coal industry is a valued part of the state economy and culture? These challenges don't stymie Jennifer Hubbard-Sanchez at Kentucky State University (KSU), for whom climate change represents the "unifying issue and opportunity of our time."

Jennifer, who has master's degrees in environmental studies and Mexican anthropological studies, is the state specialist for sustainable programs in the College of Agriculture, Food Science, and Sustainable Systems. In her projects at KSU, Jennifer combines her skills working with multicultural groups and her knowledge of climate change.

Jennifer's Climate Fellows project built on work she did through the U.S. National Oceanic and Atmospheric Administration (NOAA) Climate Stewards

Education Project (now called Planet Stewards). She developed "Climate Change 101" and conducted climate change community leader training sessions with members of KSU student groups, such as the Green Society and Minorities in Agriculture, Natural Resources, and Related Sciences. "The goal of my project really was to empower students to understand what climate change is, how we are contributing to it, and what we can do about it here locally," she said. As part of empowering students, Jennifer guided them in building partnerships with local organizations, developing communication strategies aimed at their peers, and working with the university to effect change.

Jennifer started by introducing the topic of climate change, including basic concepts like the differences between weather and climate, and using humor to make her students comfortable. "The way that I present is oftentimes pretty goofy and pretty funny. Humor always helps. Climate change is a heavy issue. And it's a heavy topic. And I think when we present it in a way that's heavy, all we're doing is scaring people and turning people away."

Once students had gained knowledge about climate change, Jennifer asked them how they wanted to address climate change in a way that raised awareness and reduced the university's carbon footprint. Jennifer's students engaged in a variety of network building, leadership, public speaking, and problem-solving activities. One activity involved riding the KSU bus to Walmart, during which they gave Climate 101 "mini" talks along the way and handed out reusable bags imprinted with tips for climate change action, thus aiming to reduce students' use of plastic bags and potentially reduce their carbon footprint (should the new bags be reused multiple times).

Jennifer used framing strategies reflected in climate change communication research, including drawing on local examples and emphasizing scientific consensus. She explained: "I think it's really hard for people to look at a picture of a polar bear and walk away and really feel like they're going to be able to make any sort of change or want to make any sort of change. I think it's important to talk about the scientific agreement about this. Climate scientists, the experts in this area, do agree that it's happening, and that there's something we can do about it."

Jennifer also had advice for climate change educators who might go home feeling hopeless about their task. "You can't win them all. And I say this from a coal state. You can't win 'em all. And that's OK. And I think that's really important to remember, to not get frustrated when you realize you just can't win them all. Channel that fear or frustration or hopelessness that I think all climate change communicators feel, channel that into something positive."

In the spring of 2016, the city of Frankfurt, Kentucky, invited Jennifer to help organize the city's annual "Reforest Frankfurt" event. Jennifer and her students hosted hundreds of volunteers, the mayor, the city commissioner, and additional KSU students. Together, they planted 2,500 trees. Although official city publications did not frame the event around climate change, for Jennifer and her students, the event was a chance to reduce the community's carbon footprint.

TABLE 12.1 Summary of Jennifer Hubbard-Sanchez's climate change programs and how they connect to concepts covered in chapters 1–10

SETTING	KENTUCKY STATE UNIVERSITY AND CITY OF FRANKFORT, KENTUCKY
Audience	Undergraduate students in student clubs
Primary program outcomes	Climate literacy Positive youth development Collective action
Climate psychology connections	Aims to reduce feelings of helplessness in the face of climate change (e.g., reduce terror management responses) by letting students choose their action projects and encouraging local actions
Climate Communication Connections	Uses positive, hopeful messaging Uses metaphors like "heat-trapping blanket" Frames climate change in terms of local and human impacts Uses humor to put students at ease
Program strategies	Pairs climate literacy activities (Climate Change 101 lectures) with opportunities for student action and leadership Fosters a community of practice among students

Summary

Jennifer structured her program to give students opportunities to lead and create action projects. She used positive language and incorporated humor in her Climate Change 101 trainings to build climate literacy while enhancing student collective action and self-efficacy, which is consistent with positive youth development outcomes (table 12.1).

Jennifer's Tip for Educators

Instead of letting the climate change blues get you down, harness your frustration into something productive and active, like a climate change action project.

13

BUILDING SOIL TO CAPTURE CARBON IN A SCHOOL GARDEN IN NEW MEXICO

The Next Generation Science Standards include explicit benchmarks for climate change education. Even as states adopt the standards, however, climate change continues to present a challenge for formal science teachers in more ways than one. Not only is it a complex, interdisciplinary topic, but teachers' own knowledge and values can make it difficult to approach.[1] Environmental educators, with their experience teaching about other so-called "wicked problems" like nuclear energy, are well prepared to meet the challenges of teaching such an interdisciplinary and difficult topic as climate change.[2] Here we tell the story of Karen Temple-Beamish, a science teacher at the Albuquerque Academy in New Mexico, who weaves climate change into her classroom, her work with school environmental clubs, and her Desert Oasis Teaching Garden.

In 2013, an irrigation system near the Albuquerque Academy broke and flooded the campus, resulting in significant erosion. Seeking a solution to the erosion and inspired by Gary Nabhan's *Growing Food in a Hotter, Drier Land*,[3] Karen Temple-Beamish used her experience as an eighth-grade science teacher and garden lover to create the Desert Oasis Teaching Garden. Karen realized that the garden "could really be a hub for teaching others in the community, faculty, staff, parents, other schools, other children at other schools, other professionals . . . how to actually grow food with less water, less resources, and at the same time build soil" to sequester carbon.

To begin building rich soil for the garden and sequestering carbon, Karen and student volunteers spread the school's compost over the garden plot and planted cover crops. The garden now encompasses two acres, with ten raised beds and a pollinator meadow. Karen and her colleagues also constructed a welcome center to provide shade and inform visitors about the garden and its practices using interpretive panels. They are planning to add an orchard.

Karen uses the garden with her students at the Albuquerque Academy, holds workshops for the local community, and hosts learn-work days. One thing that Karen has learned over the twenty years she has been teaching climate change is that "you have to give people a solution . . . something that they can participate in. And so that's how we always frame it." Karen's solutions center on carbon sequestration in the garden's soil and building garden literacy to help ensure students' and community members' resilience to climate change. Karen and her colleagues "ask people to come and not only work in the garden and get something accomplished but also learn in the process of their working."

Karen emphasizes solutions with community volunteers and classroom students. When working with volunteers to create "soil sponges" for trees, she explains how each part of the activity connects to a climate solution (figure 13.1). In her classroom, Karen tries to make learning about climate change "fun for the most part and something [students] can do something about. For example, right now, they are creating a pledge that they will try and carry out for the rest of the year. This pledge is something I asked them to do that's something they have control over." Some students are eating less meat, while others are taking shorter showers. All students are keeping a journal about their experiences and recording data. "If they're doing food, they can make a list of the food they bought that week and turn that into pounds of CO_2 emitted for the production of that food type."

Karen's tips for educators communicating about climate change are, first, make it relevant, because "these kids, they've got their hands full just growing up. And so it has

Connecting Garden Activities to Climate Change Solutions

Metaphor— Karen compares a familiar object—a sponge—to the materials the garden uses to absorb water for tree health.

Karen: And so we begin this session (with volunteers) by explaining that these trees are stressed due to lack of water. We need to be good stewards of these trees even if they aren't exactly the right trees for this particular ecosystem. We talk about the importance of trees as carbon sinks and we then go into how what we just created was like a sponge that allows the tree to keep drinking the water slowly. We've also recycled a lot of this waste material that will break down through worm activity and bacterial activity into a really good nutritious soil for the trees, so now we've brought that piece in that we're not wasting materials. So what we're doing is incorporating this talk of climate change solutions into each of the different pieces of this activity. So it's not a lecture and then let's do this activity. It's talking about each piece as we move through the solution.

Connecting to a value—Karen connects volunteer actions in the garden to the value of good stewardship.

Solutions— Karen points out that the volunteers are already engaged in climate change solutions.

FIGURE 13.1 Karen's explanation of soil sponges, in which she connects each step of the activity to climate change solutions

TABLE 13.1 Summary of Karen Temple-Beamish's climate change programs and how they connect to concepts covered in chapters 1–10

SETTING	ALBUQUERQUE ACADEMY, ALBUQUERQUE, NEW MEXICO
Audience	Middle school students General public
Primary program outcomes	Climate literacy Collective action Climate mitigation
Climate psychology connections	Uses psychologically close examples of possible solutions
Climate communication connections	Uses positive, hopeful messaging Frames climate change around solutions Makes solutions relevant by framing around students' everyday activities
Program strategies	Combines climate change lessons with opportunities for individual action at home and collective action in the garden Incorporates student reflection through keeping a journal and taking quantitative measurement of their carbon footprints Uses the garden as a means of inspiring hope and bringing together people from around the community in collective action

to connect to their lives." Second, provide audiences with "solutions that they can get their hands on and do something with." She explains that "if you hammer all those facts and figures at them, it doesn't go anywhere except maybe depress them. But if you give them something that's fun that they can do about it, then it's most effective."

In 2016, Karen participated in the Polar Trec in Alaska, where she researched carbon flux in the tundra. She has used this experience to create new climate change lessons for her students, connecting their soil-building in the garden to carbon cycles in the far-off tundra. She is also developing a project called "Children Capturing Carbon" that builds on the carbon sequestration work taking place in the garden.

Summary

Karen focuses on solutions both in the classroom and in the Desert Oasis Teaching Garden. In her classroom, students learn about individual actions they can take at home to mitigate climate change. In the garden, students and community members learn about how each garden activity ties into climate change mitigation, like building soil to sequester carbon (table 13.1).

Karen's Tip for Educators

Make climate change relevant to students' lives and solutions-based. Just using facts and figures can lead students to disengage.

14

PSYCHOLOGICAL RESILIENCE IN DENVER, COLORADO

Being fully cognizant of climate change impacts to the environment and society can be a heavy emotional burden to carry. In addition to bearing this burden in their personal lives, educators and environmental leaders work to maintain a strong emotional presence in front of their audiences. They require the psychological resilience to teach and know about climate change and yet remain positive. Here we tell the story of Maria Talero, who decided to tackle the issue of psychological resilience among environmental educators and other environmental professionals.

How do those who are concerned about climate change develop coping strategies and the psychological resilience necessary to teach and think deeply about difficult issues like climate change? (See chapter 3 for resilience definitions.) Maria Talero, a freelance climate change educator and former university philosophy professor in Denver, Colorado, decided to address this challenge. She developed Climate Courage Resilience Circles, which are small group meetings to increase members' psychological and emotional resilience through participation in climate change action. The resilience circles were so popular that Maria crowd-funded and implemented a Community Climate Courage Film and Discussion Forum. Through these experiences, Maria hoped that participants would walk away with "a strong, positive experience at the end—a feeling of excitement and energy." Her goals are to foster fellowship, psychological support, learning, and action.

Maria engages a range of audiences, from environmental professionals to concerned community members looking for ways to become involved in organized climate action. Her audiences see climate change through the lens of saving the planet and as a crisis for their grandchildren. Maria describes them as easily feeling "doom and gloom" about climate change. In her resilience circles, Maria starts by getting people on the same page about the topic of the day. As an invitation to discussion, Maria shows short videos. "There has to be a piece where we get on the same page, like an orchestra tuning instruments together." Maria chooses the videos carefully so that they present climate change as a serious issue, appeal to people's social nature, and inspire. She makes sure that immediately after the videos, people engage in some form of social interaction that avoids doom and gloom. "We want movement, we want communication, we want the learning to happen in shorter cycles. You don't want people to sit passively for too long. And you want their voices and their knowledge and their experiences to be part of what happens."

When I asked Maria about the language she uses to describe climate change, she said that she avoids catastrophe framing, tries to make her messages relevant and accessible to the audience, and frames solutions in terms of collective action. Maria explained, "If I'm going to say twenty different sentences, I want fifteen or seventeen of them to have an easy, relatable structure, and metaphor helps, examples, personal stories help, and that's so important."

When asked about the psychological barriers to climate change action, Maria, who has read widely about climate change and psychology, emphasized learned helplessness, or feeling like you have no control over a situation and thus deciding against action. She also cited environmental education's history of promoting individual-level solutions that fail to address the scale of climate change. In Maria's experience, people intuitively sense that "easy and painless" actions that people can take all by themselves, such as recycling or turning off the lights, will not solve the problem, and that "if you appeal to the single individual, you're missing out." Maria works to shift away from learned helplessness by inviting local representatives from organizations like 350 Colorado to resilience circle meetings, enabling participants to sign up that very night for collective action in their community. At film forums, "birds of a feather groups" form around interests like renewable energy collectives and become action groups that continue after the forum ends.

When asked about advice for environmental educators, Maria counseled that educators should "be really suspicious of one-size-fits-all approaches." She reiterated that recycling, a behavior perceived as a stock solution for many environmental ills, will not suffice as a climate change solution. She emphasized that the field of climate change education is in a period of "ferment and crisis" as it transitions from individual behavior changes to collective action and social behaviors, and this period means that environmental educators need to look for information and resources in new areas. Maria herself draws on a wide breadth of resources, from empirical sources like Yale's Cultural Cognition Project to books

TABLE 14.1 Summary of Maria Talero's climate change programs and how they connect to concepts covered in chapters 1–10

SETTING	VARIOUS LOCATIONS IN DENVER, COLORADO
Audience	Environmental community
Primary program outcomes	Climate literacy Self-efficacy Collective action
Climate psychology connections	Uses conversation and discussion strategically to avoid feelings of doom and gloom Provides opportunities to sign up for action projects during the program, potentially avoiding cognitive dissonance
Climate communication connections	Uses positive, hopeful messaging Frames climate change around collective solutions Frames climate change in the big picture but then provides examples of local actions participants can take Takes into account how discussion and films will impact her audience's emotions
Program strategies	Starts with an element that addresses the problem and then quickly moves to opportunities for discussion and conversation Organizes small-group discussions in film forum by putting people into different interest groups and then posing general discussion questions to the whole audience

on psychological resilience like Mary Pipher's *The Green Boat.*[1] In this book, Pipher writes about how to avoid, as Maria puts it, being "overcome by despair and how to turn toward passion as a resource, as something healing. It's not just that you're doing something about some external problem" when you take up environmental action; "you're actually helping yourself."

Summary

Maria structures her programs so participants move from knowing climate change is an issue to opportunities to socialize, discuss, build community, and, finally, to take climate change action. Her goals are to increase her audience's climate literacy and self-efficacy and to promote community and collective action. Most important, she hopes to increase audience members' psychological resilience (table 14.1).

Maria's Tip for Educators

Maria recommends that educators be wary of stock environmental solutions, like recycling, that don't actually meet the scale of the climate change problem.

PART 4 RECAP

The environmental education leaders featured in part 4 employed different strategies to achieve their outcomes. Adam Ratner at the Marine Mammal Center trained volunteers to deliver short interpretive talks that connected the center's exhibits to the issue of climate change, with the goal of inspiring community-level action and hope in visitors. Jennifer Hubbard-Sanchez trained students at Kentucky State University in basic climate change information and empowered them to create climate change outreach programs. In the Desert Oasis Teaching Garden, Karen Temple-Beamish engaged classroom students and community members in climate change mitigation, such as building soil to sequester carbon and maintaining the campus's trees with soil sponges. Maria Talero led conversation circles and film forums to bring the environmental community together in a way that encouraged learning, community, resilience, and action.

All four educators think carefully about their audiences' values, needs, and knowledge. Jennifer knew that before her students could develop climate change action plans, they would need a basic understanding of climate change, so she developed a "Climate Change 101" curriculum. She infused it with humor and levity to keep students engaged and hopeful. Maria understood that her audience came to her accepting anthropogenic climate change, so she focused on collective action and building resilience rather than literacy.

Educators also used what they knew about their audiences and environmental education practices to frame their program messages. Karen focused on collective actions to maintain her students' and volunteers' hope. Adam used explanatory metaphors drawn from NNOCCI to explain climate change. Maria avoided catastrophe frames that might induce a feeling of helplessness in her already concerned audience.

All four educators emphasized the importance of staying positive and building self- and collective efficacy among their audiences. They spoke of inspiring hope and approached climate change from a solutions-oriented perspective. Through these approaches, they fostered climate literacy and action in their communities.

CLOSING THOUGHTS

Climate change is an interdisciplinary topic that integrates perspectives from across the biophysical and social sciences. Climate change educators can benefit from a growing body of research in the social sciences. Psychological research on climate change informs how educators can assess an audience's perceptions of climate change based on that audience's identities and values and in turn shape their own communication and education strategies to achieve their program outcomes. Being familiar with psychological mechanisms such as motivated reasoning that may lead to climate denial helps educators craft programs that appeal to their particular audiences. Because research on psychological distance is inconclusive, educators may want to experiment with framing their programs with close and distant frames to find what resonates with their audiences. Communication strategies like framing, using metaphors, and using trusted messengers are fundamental tools that aid in program planning and implementation. Specifically, positive frames relevant to audiences' values, hope, and collective action help achieve climate change program outcomes.

As environmental educators strive to meet the need for climate change education, they are building on past work that focuses on individual behavior change, and on work that emphasizes collective action to better address the scale of climate change. Based on research about the tenuous relationship of knowledge to behavior, and on phenomena such as climate denial, educators are moving away from assuming information is sufficient to promote environmental action.

Instead they are taking into account audience identities, emotions, beliefs, and values, and incorporating notions such as trust, trusted messengers, and framing for collective action. In this period of uncertainty, climate change educators are trying new approaches to tackle environmental education's most critical challenge to date.

Notes

INTRODUCTION

1. Lydia Saad and Jeffrey M. Jones, "U.S. Concern about Global Warming at an Eight-Year High," Gallup.com, March 16, 2016, http://news.gallup.com/poll/190010/concern-global-warming-eight-year-high.aspx; Anthony Leiserowitz et al., *Climate Change in the American Mind: October 2017* (New Haven, CT: Yale Project on Climate Change Communication, Yale University and George Mason University, 2017), http://climatecommunication.yale.edu/publications/climate-change-american-mind-october-2017/; Gallup, "Worry about Terror Attacks in U.S. High, but Not Top Concern," Gallup.com, accessed December 19, 2017, http://news.gallup.com.proxy.library.cornell.edu/poll/190253/worry-terror-attacks-high-not-top-concern.aspx.

2. Peter Howe et al., "Geographic Variation in Opinions on Climate Change at State and Local Scales in the USA," *Nature Climate Change* 5, no. 6 (June 2015): 596–603; Riley E. Dunlap, Aaron M. McCright, and Jerrod H. Yarosh, "The Political Divide on Climate Change: Partisan Polarization Widens in the U.S.," *Environment* 58, no. 5 (September 2, 2016): 4–23, https://doi.org/10.1080/00139157.2016.1208995; Patrick J. Egan and Megan Mullin, "Climate Change: U.S. Public Opinion," *Annual Review of Political Science* 20 (2017): 209–27.

3. Marianne E. Krasny et al., *Climate Change and Environmental Education: Framing Perspectives*, Cornell University Civic Ecology Lab report, February 1, 2015.

4. Community Climate Change Fellows, *Community Climate Change Education: A Mosaic of Approaches*, ed. Marna Hauk and Elizabeth Pickett, 2016, https://naaee.org/eepro/resources/community-climate-change-education.

5. Martha C. Monroe and Annie Oxarart, eds., *Southeastern Forests and Climate Change: A Project Learning Tree Secondary Environmental Education Module* (Gainesville: University of Florida and American Forest Foundation, 2014), https://www.plt.org/southeastern-forests-and-climate-change.

6. Janet K. Swim et al., "Climate Change Education at Nature-Based Museums," *Curator: The Museum Journal* 60, no. 1 (January 1, 2017): 101–19, https://doi.org/10.1111/cura.12187; Martha C. Monroe et al., "Identifying Effective Climate Change Education Strategies: A Systematic Review of the Research," *Environmental Education Research*, 2017, 1–22.

7. Monroe et al., "Identifying Effective Climate Change Education Strategies."

8. Victoria Wibeck, "Enhancing Learning, Communication and Public Engagement about Climate Change—Some Lessons from Recent Literature," *Environmental Education Research* 20, no. 3 (2014): 387–411.

9. Anne Armstrong, "Climate Change Communication in Environmental Education: From Research to Practice" (Master's thesis, Cornell University, 2017), https://search.proquest.com/pqdtglobal/docview/1985045643/66E58850D5F34BA6PQ/1.

10. Laura Downey et al., eds., "Advancing Climate Change Environmental Education: Resources and Suggestions," EE Capacity, Cornell University, Civic Ecology Lab, North American Association of Environmental Education, 2013, http://www.eecapacity.net.

11. Yale Program on Climate Change Communication, "What Is Climate Change Communication?," accessed July 9, 2017, http://climatecommunication.yale.edu/about/what-is-climate-change-communication/.

12. North American Association for Environmental Education, "About EE and Why It Matters," NAAEE, May 19, 2015, https://naaee.org/about-us/about-ee-and-why-it-matters.

1. CLIMATE CHANGE SCIENCE

1. EPA, "Greenhouse Gas Emissions," 2017; Madeleine Rubenstein, "Emissions from the Cement Industry," *State of the Planet* (blog), May 9, 2012, http://blogs.ei.columbia.edu/2012/05/09/emissions-from-the-cement-industry/; United Nations Framework Convention on Climate Change, "Gas Emissions from Waste Disposal," 2002, http://www.grid.unep.ch/waste/download/waste4243.PDF.

2. EPA, "Greenhouse Gas Emissions."

3. Steve Graham, "John Tyndall (1820–1893)," NASA Earth Observatory, October 8, 1999; Raymond P. Sorenson, "Eunice Foote's Pioneering Research on CO_2 and Climate Warming," *Search and Discovery* article 70092, 2011.

4. John Tyndall, "On the Absorption and Radiation of Heat by Gases and Vapours, and on the Physical Connexion of Radiation, Absorption and Conduction," *Philosophical Transactions of the Royal Society of London* 151, no. 1 (1861): 28.

5. Eunice Foote, "Circumstances Affecting the Heat of the Sun's Rays," *American Journal of Science and Arts* 22 (1856): 383.

6. David A. Wells, "Heat of the Sun's Rays" (1857), in *Annual of Scientific Discovery: Or, Year-Book of Facts in Science and Art for 1857* (Boston: Gould and Lincoln, 1857), 159.

7. Foote, "Circumstances Affecting the Heat of the Sun's Rays," 383.

8. Intergovernmental Panel on Climate Change, *Climate Change 2014: Synthesis Report* (Geneva, Switzerland: IPCC, 2014).

9. Nicola Jones, "How the World Passed a Carbon Threshold and Why It Matters," *Yale Environment 360*, January 26, 2017, https://e360.yale.edu/features/how-the-world-passed-a-carbon-threshold-400ppm-and-why-it-matters.

10. World Meteorological Organization, "Greenhouse Gas Concentrations Surge to New Record," October 30, 2017, https://public.wmo.int/en/media/press-release/greenhouse-gas-concentrations-surge-new-record.

11. NASA Earth Observatory, "Global Temperature Record Broken for Third Consecutive Year," 2017, https://earthobservatory.nasa.gov/IOTD/view.php?id=89469.

12. Jerry M. Melillo, Terese Richmond, and Gary W. Yohe, *Climate Change Impacts in the United States: The Third National Climate Assessment*, U.S. Global Change Research Program, 2014, http://s3.amazonaws.com/nca2014/high/NCA3_Climate_Change_Impacts_in_the_United%20States_HighRes.pdf?download=1.

13. Jeffrey Lee, "Milankovitch Cycles," Encyclopedia of Earth, July 7, 2010, http://www.eoearth.org/view/article/154612/.

14. Rebecca Lindsey, "Global Impacts of El Niño and La Niña," NOAA, Climate.gov, February 9, 2016, https://www.climate.gov/news-features/featured-images/global-impacts-el-ni%C3%B1o-and-la-ni%C3%B1a.

15. Dennis L. Hartmann et al., "Observations: Atmosphere and Surface," in *Climate Change 2013: The Physical Science Basis; Contribution of Working Group I to the Fifth Assessment Report of the Intergovernmental Panel on Climate Change*, ed. T. F. Stocker et al. (Cambridge: Cambridge University Press, 2013); Holli Riebeek, "Global Warming," NASA Earth Observatory, June 3, 2010, http://earthobservatory.nasa.gov/Features/GlobalWarming/page4.php.

16. Ulf Buntgen et al., "Cooling and Societal Change during the Late Antique Little Ice Age from 536 to Around 660 AD," *Nature Geoscience* 9, no. 3: 231–36, https://doi.org/10.1038/ngeo2652, http://www.nature.com/ngeo/journal/v9/n3/abs/ngeo2652.html#supplementary-information; Karen Harpp, "How Do Volcanoes Affect World Climate?," *Scientific American* online, 2002, https://www.scientificamerican.com/article/how-do-volcanoes-affect-w/; Riebeek, "Global Warming"; University Corporation for Atmospheric Research, "How Volcanoes Influence Climate," 2017, https://scied.ucar.edu/short-content/how-volcanoes-influence-climate.

17. Bärbel Hönisch et al., "The Geological Record of Ocean Acidification," *Science* 335, no. 6072 (March 2012): 1058–63, https://doi.org/10.1126/science.1208277; National Climate Assessment, "Ocean Acidification," 2014, http://nca2014.globalchange.gov/report/our-changing-climate/ocean-acidification.

18. Damien Cave and Justin Gillis, "Large Sections of Australia's Great Reef Are Now Dead, Scientists Find," *New York Times*, March 15, 2017, https://www.nytimes.com/2017/03/15/science/great-barrier-reef-coral-climate-change-dieoff.html; LuAnn Dahlman, "Climate Change: Ocean Heat Content," NOAA, Climate.gov, 2015, https://www.climate.gov/news-features/understanding-climate/climate-change-ocean-heat-content.

19. Bob Berwyn, "Why Is Antarctica's Sea Ice Growing While the Arctic Melts? Scientists Have an Answer," Inside Climate News, May 31, 2016; EPA, "Climate Change Indicators: Arctic Sea Ice," 2016, https://www.epa.gov/climate-indicators/climate-change-indicators-arctic-sea-ice; NASA, Climate, "Sea Ice Extent Sinks to Record Lows at Both Poles," August 6, 2017, https://www.nasa.gov/feature/goddard/2017/sea-ice-extent-sinks-to-record-lows-at-both-poles.

20. National Research Council, "Himalayan Glaciers: Climate Change, Water Resources, and Water Security" (Washington, DC: National Academies, 2012).

21. Melillo, Richmond, and Yohe, *Climate Change Impacts*.

22. Benjamin Lowy, "When Rising Seas Transform Risk Into Certainty," *New York Times*, April 18, 2017, https://www.nytimes.com/2017/04/18/magazine/when-rising-seas-transform-risk-into-certainty.html.

23. Meghann Myers, "Rising Oceans Threaten to Submerge 128 Military Bases: Report," *Navy Times*, July 29, 2016, https://www.navytimes.com/story/military/2016/07/29/rising-oceans-threaten-submerge-18-military-bases-report/87657780/.

24. Simon Albert et al., "Sea Level Rise Swallows 5 Whole Pacific Islands," *Scientific American*, Conversation, May 9, 2016, https://www.scientificamerican.com/article/sea-level-rise-swallows-5-whole-pacific-islands/ (2016); Amanda Holpuch, "Alaskan Village Threatened by Rising Sea Levels Votes for Costly Relocation," *Guardian*, August 18, 2016, https://www.theguardian.com/us-news/2016/aug/18/alaska-shishmaref-vote-move-coastal-erosion-rising-sea-levels; Reuters, "Five Pacific Islands Lost to Rising Seas as Climate Change Hits," *Guardian*, May 10, 2016, https://www.theguardian.com/environment/2016/may/10/five-pacific-islands-lost-rising-seas-climate-change; Scott Waldman, "Maryland Island Denies Sea Level Rise, Yet Wants to Stop It," *Scientific American*, ClimateWire, June 15, 2017, https://www.scientificamerican.com/article/maryland-island-denies-sea-level-rise-yet-wants-to-stop-it/.

25. National Oceanic and Atmospheric Administration (NOAA), "Introduction to Storm Surge," n.d., https://www.nhc.noaa.gov/surge/surge_intro.pdf; NOAA, "Storm Surge Overview," n.d., http://www.nhc.noaa.gov/surge/.

26. Melillo, Richmond, and Yohe, *Climate Change Impacts*.

27. Sharon L. Harlan et al., "Neighborhood Effects on Heat Deaths: Social and Environmental Predictors of Vulnerability in Maricopa County, Arizona," *Environmental Health Perspectives* 121 (2013): 197–204, https://doi.org/10.1289/ehp.1104625.

28. Centers for Disease Control and Prevention, "Climate Change Increases the Number and Geographic Range of Disease-Carrying Insects and Ticks," n.d., https://www.cdc.gov/climateandhealth/pubs/vector-borne-disease-final_508.pdf; University Corporation for Atmospheric Research, "Climate Change and Vector-Borne Disease," 2011, https://scied.ucar.edu/longcontent/climate-change-and-vector-borne-disease; Nick Watts et al., "The *Lancet* Countdown on Health and Climate Change: From 25 Years of Inaction to a Global Transformation for Public Health," *Lancet*, 2017, https://doi.org/10.1016/S0140-6736(17)32464-9.

29. Children and Nature Network, 2018, http://www.childrenandnature.org/.

30. Susanta Kumar Padhy et al., "Mental Health Effects of Climate Change," *Indian Journal of Occupational and Environmental Medicine* 19, no. 1 (April 2015): 3–7, https://doi.org/10.4103/0019-5278.156997.

31. Intergovernmental Panel on Climate Change, 2014, *Climate Change 2014: Impacts, Adaptation, and Vulnerability; Part A: Global and Sectoral Aspects; Contribution of Working Group II to the Fifth Assessment Report of the Intergovernmental Panel on Climate Change*, ed. Christopher B. Field et al. (Cambridge: Cambridge University Press, 2014), https://www.ipcc.ch/report/ar5/.

32. Hans Joachim Schellnhuber et al., *Turn Down Heat: Why a 4°C Warmer World Must Be Avoided* (Washington, DC: World Bank, 2012), 8.

33. National Network for Ocean and Climate Change Interpretation, "The Problem with Solutions—and How to Fix It," 2017, http://climateinterpreter.org/features/problem-solutions%E2%80%94and-how-fix-it.

34. EPA, "Sources of Greenhouse Gas Emissions," 2017, https://www.epa.gov/ghgemissions/sources-greenhouse-gas-emissions.

35. Energy Star, "Energy Savings at Home," n.d., https://www.energystar.gov/index.cfm?c=heat_cool.pr_hvac.

36. Heat Smart Tompkins, "Harnessing the Efficiencies of Home Energy Performance," 2017; Cornell Cooperative Extension, "Energy Smart Community Tompkins," 2017, http://ccetompkins.org/energy/energy-smart-community-tompkins/faqs.

37. EPA, "Sources of Greenhouse Gas Emissions."

38. Community Climate Change Fellows, *Community Climate Change Education*, ed. Maura Hauk and Elizabeth Pickett (Washington, DC: North American Association for Environmental Education, 2017).

39. EPA, "Sources of Greenhouse Gas Emissions."

40. *China Daily*, "Green Platform Turns Virtual Trees into Desert Guardians," March 29, 2017, http://www.chinadaily.com.cn/regional/2017-03/29/content_28720126.htm.

41. Paul C. Stern et al., "A Value-Belief-Norm Theory of Support for Social Movements: The Case of Environmentalism," *Research in Human Ecology* 6 (1999): 81–97.

42. EPA, "Sources of Greenhouse Gas Emissions"; M. Melissa Rojas-Downing et al., "Climate Change and Livestock: Impacts, Adaptation, and Mitigation," *Climate Risk Management* 16 (January 1, 2017): 145–63, https://doi.org/10.1016/j.crm.2017.02.001.

43. Elke Stehfest et al., "Climate Benefits of Changing Diet," *Climatic Change* 95, nos. 1–2 (July 2009): 83–102.

44. New York State Department of Environmental Conservation, "A Guide to Local Action: Climate Smart Communities Certification," n.d., http://www.dec.ny.gov/energy/50845.html.

2. CLIMATE CHANGE ATTITUDES AND KNOWLEDGE

1. Robert Gifford and Reuven Sussman, "Environmental Attitudes," September 28, 2012, https://doi.org/10.1093/oxfordhb/9780199733026.013.0004.

2. P. Wesley Schultz and Florian G. Kaiser, "Promoting Pro-Environmental Behavior," *The Oxford Handbook of Environmental and Conservation Psychology*, September 28, 2012, https://doi.org/10.1093/oxfordhb/9780199733026.013.0029.

3. Gifford and Sussman.

4. Thomas A. Heberlein and J. Stanley Black, "Attitudinal Specificity and the Prediction of Behavior in a Field Setting," *Journal of Personality and Social Psychology* 33, no. 4 (1976): 474.

5. Tien Ming Lee et al., "Predictors of Public Climate Change Awareness and Risk Perception around the World," *Nature Climate Change* 5, no. 11 (November 2015): 1014, https://doi.org/10.1038/nclimate2728.

6. Richard Wike, "What the World Thinks about Climate Change in 7 Charts," *Pew Research Center* (blog), April 18, 2016, http://www.pewresearch.org/fact-tank/2016/04/18/what-the-world-thinks-about-climate-change-in-7-charts/.

7. Chelsea Combest-Friedman, Patrick Christie, and Edward Miles, "Household Perceptions of Coastal Hazards and Climate Change in the Central Philippines," *Journal of Environmental Management* 112, no. 15 (2012): 137–48.

8. Leiserowitz et al., "Climate Change in the American Mind: October 2017," Yale Project on Climate Change Communication, Yale University and George Mason University, 2017, http://climatecommunication.yale.edu/publications/climate-change-american-mind-october-2017/.

9. Leiserowitz et al.

10. Edward Maibach et al., "Identifying Like-Minded Audiences for Global Warming Public Engagement Campaigns: An Audience Segmentation Analysis and Tool Development," *PLOS ONE*, no. 6 (2011): 3, doi:10.1371/journal.pone.0017571.

11. Aaron M. McCright and Riley E. Dunlap, "The Politicization of Climate Change and Polarization in the American Public's Views of Global Warming, 2001–2010," *Sociological Quarterly* 52, no. 2 (2011): 155–94.

12. Yale Program on Climate Change Communication, "Americans' Knowledge of Climate Change," accessed March 4, 2016, http://climatecommunication.yale.edu/publications/americans-knowledge-of-climate-change/.

13. Yale Program on Climate Change Communication, "Americans' Knowledge of Climate Change."

14. Eric Plutzer et al., "Climate Confusion among US Teachers," *Science* 351 (6274): 664–65.

15. Jing Shi et al., "Knowledge as a Driver of Public Perceptions about Climate Change Reassessed," *Nature Climate Change* 6, no. 8 (April 25, 2016): 759–62, https://doi.org/10.1038/nclimate2997.

16. Sander L. van der Linden, Anthony A. Leiserowitz, Geoffrey D. Feinberg, and Edward W. Maibach, "The Scientific Consensus on Climate Change as a Gateway Belief: Experimental Evidence," *PLOS ONE* 10, no. 2 (February 25, 2015): e0118489, doi:10.1371/journal.pone.0118489.

17. Anthony Leiserowitz, Nicholas Smith, and Jennifer R. Marlon, "American Teens' Knowledge of Climate Change," Yale Project on Climate Change Communication, 2011, http://oceanservice.noaa.gov/education/pd/climate/teachingclimate/am_teens_knowledge_of_climate_change.pdf.

18. Kathryn T. Stevenson, M. Nils Peterson, and Howard D. Bondell, "The Influence of Personal Beliefs, Friends, and Family in Building Climate Change Concern among Adolescents," *Environmental Education Research*, 2016, 1–14.

19. CUSP, "Climate Urban Systems Partnership (CUSP), Climate Change Education Kit Development," accessed January 20, 2018, http://pittsburgh.cuspproject.org/wp-content/uploads/sites/2/2017/03/Get-to-the-Game-kit-development-Document.pdf.

20. Dipesh Chakrabarty, "The Politics of Climate Change Is More than the Politics of Capitalism," *Theory, Culture & Society* 34, no. 2–3 (May 1, 2017): 35, https://doi.org/10.1177/0263276417690236.

21. Yale Program on Climate Change Communication, "Global Warming's Six Indias," *Yale Program on Climate Change Communication* (blog), accessed January 9, 2018, http://climatecommunication.yale.edu/publications/global-warmings-six-indias/.

3. CLIMATE CHANGE EDUCATION OUTCOMES

1. Adapted from Alex Russ, ed., *Measuring Environmental Education Outcomes* (Ithaca, NY, and Washington, DC: EECapacity Project, Cornell University Civic Ecology Lab and NAAEE, 2014).

2. Joe E. Heimlich and Nicole M. Ardoin, "Understanding Behavior to Understand Behavior Change: A Literature Review," *Environmental Education Research* 14, no. 3 (June 1, 2008): 215–37, https://doi.org/10.1080/13504620802148881; Anja Kollmuss and Julian Agyeman, "Mind the Gap: Why Do People Act Environmentally and What Are the Barriers to Pro-environmental Behavior?," *Environmental Education Research* 8, no. 3 (2002): 239–60, https://doi.org/10.1080/13504620220145401.

3. Anita Pugliese and Julie Ray, "Awareness of Climate Change and Threat Vary by Region," Gallup.com, December 11, 2009, http://www.gallup.com/poll/124652/Awareness-Climate-Change-Threat-Vary-Region.aspx.

4. Zarrintaj Aminrad, S. Z. B. S. Zakariya, Abdul Samad Hadi, and Mahyar Sakari, "Relationship between Awareness, Knowledge, and Attitudes towards Environmental Education among Secondary School Students in Malaysia," 2013, http://citeseerx.ist.psu.edu/viewdoc/download?doi=10.1.1.388.818&rep=rep1&type=pdf.

5. Heimlich and Ardoin, "Understanding Behavior to Understand Behavior Change."

6. *Climate Literacy: The Essential Principles of Climate Science; A Guide for Individuals and Communities*, U.S. Global Change Research Program, 2009, http://cpo.noaa.gov/sites/cpo/Documents/pdf/ClimateLiteracyPoster-8_5x11_Final4-11LR.pdf.

7. *Climate Literacy: The Essential Principles of Climate Science.*

8. *Climate Change Education: Goals, Audiences, and Strategies: A Workshop Summary* (Washington, DC: National Academies, 2011), http://www.nap.edu/catalog/13224.

9. Karen S. Hollweg, Jason Taylor, Rodger W. Bybee, Thomas J. Marcinkowski, William C. McBeth, and Pablo Zoido, *Developing a Framework for Assessing Environmental Literacy: Executive Summary*, North American Association of Environmental Education, 2011, https://naaee.org/sites/default/files/envliteracyexesummary.pdf.

10. Eliot R. Smith and Diane M. Mackie, "Attitudes and Attitude Change," in *Social Psychology*, 3rd ed. (New York: Psychology Press, 2007), 228–67.

11. Thomas A. Heberlein, *Navigating Environmental Attitudes* (New York: Oxford University Press, 2012).

12. Marc J. Stern, Robert B. Powell, and Nicole M. Ardoin, "What Difference Does It Make? Assessing Outcomes from Participation in a Residential Environmental Education Program," *Journal of Environmental Education* 39, no. 4 (July 1, 2008): 31–43, doi:10.3200/JOEE.39.4.31-43.

13. Shalom Schwartz, "Are There Universal Aspects in the Structure and Contents of Human Values?," *Journal of Social Issues* 50, no. 4 (1994): 19–45.

14. Crispin Sartwell, "Knowledge Is Merely True Belief," *American Philosophical Quarterly* 28, no. 2 (1991): 157–65.

15. Phoebe Ellsworth and Klaus R. Scherer, "Appraisal Processes in Emotion," in *Handbook of Affective Sciences*, ed. Richard J. Davidson, Klaus R. Scherer, and H. Hill Goldsmith (Oxford: Oxford University Press, 2003), 572–95.

16. C. R Snyder et al., "The Will and the Ways: Development and Validation of an Individual-Differences Measure of Hope," *Journal of Personality and Social Psychology* 60, no. 4 (1991): 570–85.

17. Maria Ojala, "Hope and Climate Change: The Importance of Hope for Environmental Engagement among Young People," *Environmental Education Research* 18, no. 5 (2012): 625–42; Kathryn T. Stevenson and Nils Peterson, "Motivating Action through Fostering Climate Change Hope and Concern and Avoiding Despair among Adolescents," *Sustainability* 8, no. 1 (January 2016): 6, https://doi.org/10.3390/su8010006.

18. Robert A. C. Ruiter, Charles Abraham, and Gerjo Kok, "Scary Warnings and Rational Precautions: A Review of the Psychology of Fear Appeals," *Psychology & Health* 16, no. 6 (November 2001): 613.

19. Harold R. Hungerford and Trudi L. Volk, "Changing Learner Behavior through Environmental Education," *Journal of Environmental Education* 21, no. 3 (March 1, 1990): 8–21, https://doi.org/10.1080/00958964.1990.10753743.

20. Albert Bandura, *Social Learning Theory* (Englewood Cliffs, NJ: Prentice-Hall, 1977).

21. Tania M. Schusler and Marianne E. Krasny, "Environmental Action and Positive Youth Development," in *Across the Spectrum*, 2nd ed. (North American Association of Environmental Education, 2015), 107–29, http://www.naaee.net/sites/default/files/publications/eebook/AcrosstheSpectrum_SU15_final_spreads.pdf.

22. Schusler and Krasny, "Environmental Action."

23. Krasny, Marianne E., Leigh Kalbacker, Richard C. Stedman, and Alex Russ, "Measuring Social Capital among Youth: Applications in Environmental Education," *Environmental Education Research* 21, no. 1 (2015): 1–23.

24. Marc J. Stern and Kimberly J. Coleman, "The Multidimensionality of Trust: Applications in Collaborative Natural Resource Management," *Society and Natural Resources* 28 (2015): 117–32.

25. Nicole M. Ardoin, Maria L. DiGiano, and Kathleen O'Connor, "The Development of Trust in Residential Environmental Education Programs," *Environmental Education Research* 23, no. 9 (February 17, 2016): 1335–1355, doi:10.1080/13504622.2016.1144176.

26. Matthew C. Nisbet and John E. Kotcher, "A Two-Step Flow of Influence? Opinion Leader Campaigns on Climate Change," *Science Communication* 30, no. 3 (January 2009): 328–54.

27. Martha L. McCoy and Patrick Scully, "Deliberative Dialogue to Expand Civic Engagement: What Kind of Talk Does Democracy Need?," *National Civic Review* 91, no. 2 (2002): 117–35.

28. Christopher Latimer and Karen Hempson, "Using Deliberation in the Classroom: A Teaching Pedagogy to Enhance Student Knowledge, Opinion Formation, and Civic Engagement," *Journal of Political Science Education* 8 (2012): 372–88; Michele Archie, Scott London, and Bora Simmons, "Climate Choices: How Should We Meet the Challenges of a Warming Planet?," National Issues Forum Institute, 2016, https://www.nifi.org/sites/default/files/product-downloads/Climate%20Choices.pdf.

29. Robert J. Sampson, Stephen W. Raudenbush, and Felton Earls, "Neighborhoods and Violent Crime: A Multilevel Study of Collective Efficacy," *Science* 277, no. 5328 (August 15, 1997): 918–24, https://doi.org/10.1126/science.277.5328.918.

30. Louise Chawla and Debra Flanders Cushing, "Education for Strategic Environmental Behavior," *Environmental Education Research* 13, no. 4 (2007): 442, https://doi.org/10.1080/13504620701581539.

31. Chawla and Cushing, "Education for Strategic Environmental Behavior."

32. John Scott and Gordon Marshall, *A Dictionary of Sociology*, 3rd ed. rev. (Oxford: Oxford University Press, 2009), http://www.oxfordreference.com/view/10.1093/acref/9780199533008.001.0001/acref-9780199533008-e-312.

33. Linda Camino and Shepherd Zeldin, "From Periphery to Center: Pathways for Youth Civic Engagement in the Day-to-Day Life of Communities," *Applied Developmental Science* 6, no. 4 (2002): 214.

34. NASA, Global Climate Change: Vital Signs of the Planet, "Responding to Climate Change," July 7, 2016, http://climate.nasa.gov/solutions/adaptation-mitigation.

35. NASA, Global Climate Change: Vital Signs of the Planet, "Responding to Climate Change."

36. Marianne E. Krasny and Bryce DuBois, "Climate Adaptation Education: Embracing Reality or Abandoning Environmental Values," *Environmental Education Research*, June 16, 2016, 1–12, https://doi.org/10.1080/13504622.2016.1196345.

37. Bryce Dubois and Marianne E. Krasny, "Educating with Resilience in Mind: Addressing Climate Change in Post-Sandy New York City," *Journal of Environmental Education* 47, no. 4 (April 29, 2016): 1–16, https://doi.org/10.1080/00958964.2016.1167004.

38. Ann Masten and Jelena Obradovic, "Disaster Preparation and Recovery: Lessons from Research on Resilience in Human Development," *Ecology and Society* 13, no. 1 (2008).

39. Community and Regional Resilience Institute (CARRI), *Definitions of Community Resilience: An Analysis* (Washington, DC: CARRI, 2013).

40. C. S. Holling, "Resilience and Stability of Ecological Systems," *Annual Review of Ecology, Evolution, and Systematics* 4 (1973): 1–23.

41. Brian Walker and David Salt, *Resilience Thinking: Sustaining Ecosystems and People in a Changing World* (Island Press, 2012).

4. CLIMATE CHANGE EDUCATION VIGNETTES

1. National Oceanic and Atmospheric Administration, "Living Shoreline," NOAA Shoreline Glossary, accessed October 25, 2017, https://shoreline.noaa.gov/glossary.html.

2. Travis M. Andrews, "Trump Calls Mayor of Shrinking Chesapeake Island and Tells Him Not to Worry about It," *Washington Post*, June 14, 2017, https://www.washingtonpost.com/news/morning-mix/wp/2017/06/14/trump-calls-mayor-of-shrinking-chesapeake-island-and-tells-him-not-to-worry-about-it/?utm_term=.bfcc3c6e0df8.

3. "HS-ESS3 Earth and Human Activity," Next Generation Science Standards, accessed March 7, 2016, http://www.nextgenscience.org/dci-arrangement/hs-ess3-earth-and-human-activity; Kentucky Department of Education, "Next Generation Science Standards (NGSS)," accessed December 30, 2017, https://education.ky.gov/curriculum/conpro/science/Pages/Next-Generation-Science-Standards.aspx.

4. Martha C. Monroe and Annie Oxarart, eds., *Southeastern Forests and Climate Change: A Project Learning Tree Secondary Environmental Education Module* (Gainesville: University of Florida and American Forest Foundation, 2014), https://www.plt.org/southeastern-forests-and-climate-change.

5. *National Geographic*, "The Carbon Bathtub," December 2009, http://ngm.nationalgeographic.com/big-idea/05/carbon-bath.

6. Christopher Latimer and Karen Hempson, "Using Deliberation in the Classroom: A Teaching Pedagogy to Enhance Student Knowledge, Opinion Formation, and Civic Engagement," *Journal of Political Science Education* 8 (2012): 372–88.

PART 1. RECAP

1. Yale Program on Climate Change Communication, "Yale Climate Opinion Maps—U.S. 2016," accessed January 3, 2018, http://climatecommunication.yale.edu/visualizations-data/ycom-us-2016/.

PART 2. THE PSYCHOLOGY OF CLIMATE CHANGE

1. Robert Gifford, "The Dragons of Inaction: Psychological Barriers That Limit Climate Change Mitigation and Adaptation," *American Psychologist* 66, no. 4 (May–June 2011): 290–302, doi:10.1037/a0023566.

2. Kristen Kunkle and Martha C. Monroe, "Misconceptions and Psychological Mechanisms of Climate Change Communication," in *Across the Spectrum: Resources for Environmental Educators*, ed. Martha Monroe and Marianne Krasny, 3rd ed. (North American Association of Environmental Education, 2016).

5. IDENTITY

1. Kristen Kunkle and Martha Monroe, "Misconceptions and Psychological Mechanisms of Climate Change Communication," in *Across the Spectrum: Resources for Environmental Educators*, ed. Martha Monroe and Marianne Krasny, 3rd ed. (North American Association of Environmental Education, 2016).

2. P. Sol Hart and Erik C. Nisbet, "Boomerang Effects in Science Communication: How Motivated Reasoning and Identity Cues Amplify Opinion Polarization about Climate Mitigation Policies," *Communication Research* 39, no. 6 (December 1, 2012): 701–23, https://doi.org/10.1177/0093650211416646.

3. Anja Kollmuss and Julian Agyeman, "Mind the Gap: Why Do People Act Environmentally and What Are the Barriers to Pro-environmental Behavior?," *Environmental Education Research* 8, no. 3 (2002): 239–60, https://doi.org/10.1080/13504620220145401; David Ockwell, Lorraine Whitmarsh, and Saffron O'Neill, "Reorienting Climate Change Communication for Effective Mitigation: Forcing People to Be Green or Fostering Grass-Roots Engagement?," *Science Communication*, January 7, 2009, https://doi.org/10.1177/1075547008328969.

4. Susan D. Clayton, *The Oxford Handbook of Environmental and Conservation Psychology* (New York: Oxford University Press, 2012), 165.

5. Clayton, 167.

6. Sarah Riggs Stapleton, "Environmental Identity Development through Social Interactions, Action, and Recognition," *Journal of Environmental Education* 46, no. 2 (2015): 94–113.

7. Kelly S. Fielding and Matthew J. Hornsey, "A Social Identity Analysis of Climate Change and Environmental Attitudes and Behaviors: Insights and Opportunities," *Frontiers in Psychology* 7, no. 121 (2016), https://www.ncbi.nlm.nih.gov/pmc/articles/PMC4749709/.

8. Daphna Oyserman, "Identity-Based Motivation: Implications for Action-Readiness, Procedural-Readiness, and Consumer Behavior," *Journal of Consumer Psychology* 19, no. 3 (July 2009): 250–60, https://doi.org/10.1016/j.jcps.2009.05.008.

9. Ziva Kunda, "The Case for Motivated Reasoning," *Psychological Bulletin* 108, no. 3 (1990): 480.

10. Kunda.

11. Jonathon P. Schuldt and Sungjong Roh, "Media Frames and Cognitive Accessibility: What Do 'Global Warming' and 'Climate Change' Evoke in Partisan Minds?," *Environmental Communication* 8, no. 4 (October 2, 2014): 529–48, https://doi.org/10.1080/17524032.2014.909510.

12. Dan M. Kahan, Ellen Peters, Maggie Wittlin, Paul Slovic, Lisa Larrimore Ouellette, Donald Braman, and Gregory N. Mandel, "The Polarizing Impact of Science Literacy and Numeracy on Perceived Climate Change Risks," *Nature Climate Change* 2 (October 2012): 732–35.

13. "The Psychology of Climate Change Communication: A Guide for Scientists, Journalists, Educators, Political Aides, and the Interested Public," Center for Research on Environmental Decisions, 2009, http://guide.cred.columbia.edu/pdfs/CREDguide_full-res.pdf.

14. Ellen Peters, "Why We Don't Believe Science: A Perspective from Decision Psychology," webinar, Global Change Local Impact: OSU Climate Change Webinar Series, Ohio State University, February 11, 2016.

15. Peters, "Why We Don't Believe Science"; Ariel Malka, Jon A. Krosnick, and Gary Langer, "The Association of Knowledge with Concern about Global Warming: Trusted Information Sources Shape Public Thinking," *Risk Analysis* 29, no. 5 (May 1, 2009): 633–47, doi:10.1111/j.1539-6924.2009.01220.x.

16. Hart and Nisbet, "Boomerang Effects."

17. Lawrence C. Hamilton, "Education, Politics and Opinions about Climate Change: Evidence for Interactive Effects," *Climatic Change* 103 (January 1, 2011): 231–42, doi:10.1007/s10584-010-9957-8.

18. Anthony N. Washburn and Linda J. Skitka, "Science Denial across the Political Divide: Liberals and Conservatives Are Similarly Motivated to Deny Attitude-Inconsistent Science," *Social Psychological and Personality Science*, 2017.

19. Matthew J. Hornsey et al., "Meta-Analyses of the Determinants and Outcomes of Belief in Climate Change," *Nature Climate Change* 6 (2016): 622–26.

20. Ana-Maria Bliuc et al., "Public Division about Climate Change Rooted in Conflicting Socio-Political Identities," *Nature Climate Change* 5, no. 3 (March 2015): 226–29, doi:10.1038/nclimate2507.

21. Jonathon P. Schuldt and Adam R. Pearson, "The Role of Race and Ethnicity in Climate Change Polarization: Evidence from a U.S. National Survey Experiment," *Climatic Change* 136, nos. 3–4 (2016): 1–11, doi:10.1007/s10584-016-1631-3.

22. Michael Paolisso et al., "Climate Change, Justice, and Adaptation among African American Communities in the Chesapeake Bay Region," *Weather, Climate, and Society* 4, no. 1 (January 1, 2012): 34–47, https://doi.org/10.1175/WCAS-D-11-00039.1.

23. Peter Howe et al., "Geographic Variation in Opinions on Climate Change at State and Local Scales in the USA," *Nature Climate Change* 5, no. 6 (June 2015): 596–603.

24. Dan M. Kahan, Donald Braman, John Gastil, Paul Slovic, and C. K. Mertz, "Culture and Identity-Protective Cognition: Explaining the White-Male Effect in Risk Perception," *Journal of Empirical Legal Studies* 4, no. 3 (2007): 465–505; Aaron M. McCright and Riley E. Dunlap, "Cool Dudes: The Denial of Climate Change among Conservative White Males in the United States," *Global Environmental Change* 21, no. 4 (October 2011): 1163–72, doi:10.1016/j.gloenvcha.2011.06.003.

25. Adam R. Pearson, Jonathon P. Schuldt, and Rainer Romero-Canyas, "Social Climate Science: A New Vista for Psychological Science," *Perspectives on Psychological Science*, 2016, http://research.pomona.edu/sci/files/2016/06/PearsonSchuldtRomeroCanyas2016PPS-Social-Climate-Science.pdf.

26. Adam R. Pearson et al., "Race, Class, Gender and Climate Change Communication," *Oxford Research Encyclopedias: Climate Science* (Oxford University Press, April 26, 2017), http://climatescience.oxfordre.com/view/10.1093/acrefore/9780190228620.001.0001/acrefore-9780190228620-e-412.

27. Dorceta E. Taylor, *The State of Diversity in Environmental Organizations: Mainstream NGOs, Foundations, Government Agencies*, Green 2.0, 2014, https://www.diversegreen.org/wp-content/uploads/2015/10/FullReport_Green2.0_FINAL.pdf.

28. Efrat Eizenberg, *From the Ground Up: Community Gardens in New York City and the Politics of Spatial Transformation* (New York: Routledge, 2016); Taylor, *State of Diversity in Environmental Organizations.*

29. Katie Worth, "Climate Change Skeptic Group Seeks to Influence 200,000 Teachers," *Frontline*, accessed November 8, 2017, https://www.pbs.org/wgbh/frontline/article/climate-change-skeptic-group-seeks-to-influence-200000-teachers/.

30. Amy Harmon, "Climate Science Meets a Stubborn Obstacle: Students," *New York Times*, June 4, 2017, https://www.nytimes.com/2017/06/04/us/education-climate-change-science-class-students.html.

31. Stapleton, "Environmental Identity Development"; Carie Green, Darius Kalvaitis, and Anneliese Worster, "Recontextualizing Psychosocial Development in Young Children: A Model of Environmental Identity Development," *Environmental Education Research*, August 14, 2015, 1–24, https://doi.org/10.1080/13504622.2015.1072136; Corrie Colvin Williams and Louise Chawla, "Environmental Identity Formation in Nonformal Environmental Education Programs," *Environmental Education Research* 22, no. 7 (2015): 978–1001.

32. Kathryn T. Stevenson, M. Nils Peterson, Howard D. Bondell, Susan E. Moore, and Sarah J. Carrier, "Overcoming Skepticism with Education: Interacting Influences of Worldview and Climate Change Knowledge on Perceived Climate Change Risk among Adolescents," *Climatic Change* 126, nos. 3–4 (August 2014): 293–304, doi:10.1007/s10584-014-1228-7.

33. Kathryn T. Stevenson et al., "How Emotion Trumps Logic in Climate Change Risk Perception: Exploring the Affective Heuristic among Wildlife Science Students," *Human Dimensions of Wildlife* 20, no. 6 (2015): 501–513.

34. Kelly S. Fielding and Matthew J. Hornsey, "A Social Identity Analysis of Climate Change and Environmental Attitudes and Behaviors: Insights and Opportunities," *Personality and Social Psychology*, 2016, 121, https://doi.org/10.3389/fpsyg.2016.00121.

35. Bliuc et al., "Public Division about Climate Change."

6. PSYCHOLOGICAL DISTANCE

1. Sam H. Ham, *Environmental Interpretation: A Practical Guide for People with Big Ideas and Small Budgets* (Golden, CO: North American, 1992); Martha C. Monroe et al., "Identifying Effective Climate Change Education Strategies: A Systematic Review of the Research," *Environmental Education Research*, 2017, 1–22.

2. Martha C. Monroe and Annie Oxarart, eds., *Southeastern Forests and Climate Change: A Project Learning Tree Secondary Environmental Education Module* (Gainesville: University of Florida and American Forest Foundation, 2014), http://sfrc.ufl.edu/extension/ee/climate/.

3. Alexa Spence, Wouter Poortinga, and Nick Pidgeon, "The Psychological Distance of Climate Change," *Risk Analysis* 32, no. 6 (June 1, 2012): 957–72, https://doi.org/10.1111/j.1539-6924.2011.01695.x.

4. Spence, Poortinga, and Pidgeon.

5. Irene Lorenzoni, Sophie Nicholson-Cole, and Lorraine Whitmarsh, "Barriers Perceived to Engaging with Climate Change among the UK Public and Their Policy Implications," *Global Environmental Change* 17, no. 3 (2007): 446.

6. Yaacov Trope, Nira Liberman, and Cheryl Wakslak, "Construal Levels and Psychological Distance: Effects on Representation, Prediction, Evaluation, and Behavior," *Journal of Consumer Psychology: The Official Journal of the Society for Consumer Psychology* 17, no. 2 (2007): 83.

7. Trope, Liberman, and Wakslak.

8. Adrian Brügger, Suraje Dessai, Patrick Devine-Wright, Thomas A. Morton, and Nicholas F. Pigeon, "Psychological Responses to the Proximity of Climate Change," *Nature Climate Change* 5, no. 12 (December 2015): 1031–37, doi:10.1038/nclimate2760.

9. Anthony A. Leiserowitz, "American Risk Perceptions: Is Climate Change Dangerous?," *Risk Analysis* 25, no. 6 (2005): 1433–1442, doi:10.1111/j.1540-6261.2005.00690.x.

10. Alexa Spence and Nick Pidgeon, "Framing and Communicating Climate Change: The Effects of Distance and Outcome Frame Manipulations," *Global Environmental Change* 20, no. 4 (2010): 656–67.

11. Leila Scannell and Robert Gifford, "Personally Relevant Climate Change: The Role of Place Attachment and Local versus Global Message Framing in Engagement," *Environment and Behavior* 45, no. 1 (2013): 60–85.

12. Hang Lu and Jonathon P. Schuldt, "Compassion for Climate Change Victims and Support for Mitigation Policy," *Journal of Environmental Psychology* 45 (2016): 192–200.

13. Lu and Schuldt.

14. Laura N. Rickard, Z. Janet Yang, and Jonathon P. Schuldt, "Here and Now, There and Then: How 'Departure Dates' Influence Climate Change Engagement," *Global Environmental Change* 38 (May 2016): 97–107, https://doi.org/10.1016/j.gloenvcha.2016.03.003.

15. Adrian Brügger, Thomas A. Morton, and Suraje Dessai, "'Proximising' Climate Change Reconsidered: A Construal Level Theory Perspective," *Journal of Environmental Psychology* 46, no. Supplement C (June 1, 2016): 125–42, https://doi.org/10.1016/j.jenvp.2016.04.004.

16. Lu and Schuldt, "Compassion for Climate Change Victims."

17. Rickard, Yang, and Schuldt, "Here and Now, There and Then."

7. OTHER PSYCHOLOGICAL THEORIES

1. Janet K. Swim and John Fraser, "Fostering Hope in Climate Change Educators," *Journal of Museum Education* 38, no. 3 (2013): 286–97.

2. Janis L. Dickinson, "The People Paradox: Self-Esteem Striving, Immortality Ideologies, and Human Response to Climate Change," *Ecology and Society* 14, no. 1 (2009): 34.

3. Mark J. Landau, Sheldon Solomon, Jeff Greenberg, Tom Pyszczynski, Jamie Arndt, Claude H. Miller, Daniel M. Oglive, and Alison Cook, "Deliver Us from Evil: The Effects of Mortality Salience and Reminders of 9/11 on Support for President George W. Bush," *Personality and Social Psychology Bulletin* 30, no. 9 (2004): 1136–1150.

4. Isabella Uhl et al., "Undesirable Effects of Threatening Climate Change Information: A Cross-Cultural Study," *Group Processes and Intergroup Relations*, 2017, 1–17.

5. Richard Wilk, "Consumption, Human Needs, and Global Environmental Change," *Global Environmental Change* 12, no. 1 (April 2002): 5–13, doi:10.1016/S0959-3780(01)00028-0.

6. Julie S. Johnson-Pynn and Laura R. Johnson, "Successes and Challenges in East African Conservation Education," *Journal of Environmental Education* 36, no. 2 (2005): 25–39.

7. Leon Festinger, "Cognitive Dissonance," *Scientific American* 207 (1962): 93–102, doi:10.1037/10318-001.

8. Per Espen Stoknes, *What We Think About When We Try Not to Think About Global Warming: Toward a New Psychology of Climate Action* (Chelsea Green, 2015), 64.

PART 2 RECAP

1. National Audubon Society, *Tools of Engagement: A Toolkit for Engaging People in Conservation*, ed. Judy Braus (National Audubon Society, 2011).

2. Bora Simmons, Ed McCrea, Andrea Shotkin, Drew Burnett, Kathy McGlauflin, Richard Osorio, Celeste Prussia, Andy Spencer, and Brenda Weiser, *Nonformal Environmental Education Programs: Guidelines for Excellence*, North American Association of Environmental Education, 2009, https://naaee.org/sites/default/files/nonformalgl.pdf.

8. FRAMING CLIMATE CHANGE

1. Dennis Chong and James N. Druckman, "A Theory of Framing and Opinion Formation in Competitive Elite Environments," *Journal of Communication* 57, no. 1 (2007): 99–118.

2. Matthew C. Nisbet, "Communicating Climate Change: Why Frames Matter for Public Engagement," *Environment: Science and Policy for Sustainable Development* 51, no. 2 (2009): 12–23.

3. Nisbet.

4. Jonathon P. Schuldt and Sungjong Roh, "Media Frames and Cognitive Accessibility: What Do 'Global Warming' and 'Climate Change' Evoke in Partisan Minds?," *Environmental Communication* 8, no. 4 (2014): 530, http://doi.org/10.1080/17524032.2014.909510.

5. EPA, "Learn the Basics," accessed April 4, 2016, https://www3.epa.gov/climate change//kids/index.html.

6. "What's the Deal?," *Alliance for Climate Education*, March 17, 2015, https://acespace.org/the-deal.

7. George Lakoff, "Why It Matters How We Frame the Environment," *Environmental Communication* 4, no. 1 (2010): 70–81.

8. James N. Druckman, "The Implications of Framing Effects for Citizen Competence," Political *Behavior* 23, no. 3 (2001): 225–56.

9. Irwin P. Levin and Gary J. Gaeth, "How Consumers Are Affected by the Framing of Attribute Information before and after Consuming the Product," *Journal of Consumer Research* 15, no. 3 (December 1, 1988): 374–78, https://doi.org/10.1086/209174.

10. Thomas A. Morton et al., "The Future That May (or May Not) Come: How Framing Changes Responses to Uncertainty in Climate Change Communications," *Global Environmental Change* 21, no. 1 (February 2011): 103–9, https://doi.org/10.1016/j.gloenvcha.2010.09.013.

11. Sander L. van der Linden, Anthony A. Leiserowitz, Geoffrey D. Feinberg, and Edward W. Maibach, "The Scientific Consensus on Climate Change as a Gateway Belief: Experimental Evidence," *PLOS ONE* 10, no. 2 (February 25, 2015): e0118489, https://doi.org/10.1371/journal.pone.0118489.

12. Leona Yi-Fan Su, Heather Akin, Dominique Brossard, Dietram A. Scheufele, and Michael A. Xenos, "Science News Consumption Patterns and Their Implications for Public Understanding of Science," *Journalism & Mass Communication Quarterly* 92, no. 3 (2015): 597–616.

13. P. Sol Hart and Lauren Feldman, "Threat without Efficacy? Climate Change on U.S. Network News," *Science Communication* 36, no. 3 (February 2014): 325–51, https://doi.org/10.1177/1075547013520239.

14. Lisa Antilla, "Climate of Skepticism: US Newspaper Coverage of the Science of Climate Change," *Global Environmental Change* 15 (2005): 338–52.

15. Michael Hiltzik, "Climate Change Will Be an Economic Disaster for Rich and Poor, New Study Says," *Los Angeles Times*, March 18, 2016, http://www.latimes.com/business/hiltzik/la-fi-mh-climate-change-economic-disaster-20151026-column.html.

16. Bjorn Lomborg, "Gambling the World Economy on Climate," *Wall Street Journal*, November 17, 2015, Opinion, http://www.wsj.com/articles/gambling-the-world-economy-on-climate-1447719037.

17. Steven E. Koonin, "Climate Science Is Not Settled," *Wall Street Journal*, September 19, 2014, Life and Style, http://www.wsj.com/articles/climate-science-is-not-settled-1411143565.

18. Eric Zuesse, "Climate Catastrophe Will Hit Tropics around 2020, Rest of World around 2047, Study Says," *Huffpost Green*, October 14, 2013, http://www.huffingtonpost.com/eric-zuesse/climate-catastrophe-to-hi_b_4089746.html.

19. Robb Willer, "Is the Environment a Moral Cause?," *New York Times*, February 27, 2015, http://www.nytimes.com/2015/03/01/opinion/sunday/is-the-environment-a-moral-cause.html.

20. Christine Otieno et al., "Informing about Climate Change and Invasive Species: How the Presentation of Information Affects Perception of Risk, Emotions, and Learning," *Environmental Education Research* 20, no. 5 (September 3, 2014): 612–38, https://doi.org/10.1080/13504622.2013.833589.

21. Anne Armstrong, "Climate Change Communication in Environmental Education: From Research to Practice" (master's thesis, Cornell University, 2017), https://search.proquest.com/pqdtglobal/docview/1985045643/66E58850D5F34BA6PQ/1.

22. K. C. Busch, "Polar Bears or People? Exploring Ways in Which Teachers Frame Climate Change in the Classroom," *International Journal of Science Education, Part B* 6, no. 2 (April 2, 2016): 137–65, https://doi.org/10.1080/21548455.2015.1027320.

23. FrameWorks Institute, accessed February 24, 2016, http://www.frameworksinstitute.org/.

24. EPA, "Learn the Basics."

25. Armstrong, "Climate Change Communication."

26. Laura Rose and Lisa Ayers Lawrence, "Coral Bleaching: A White Hot Problem," Bridge: An Ocean of Free Teacher-Approved Marine Education Resources, accessed July 5, 2016, http://www2.vims.edu/bridge/DATA.cfm?Bridge_Location=archive0406.html.

27. "A Subsistence Culture Impacted by Climate Change," PBS LearningMedia, accessed April 4, 2016, http://www.pbslearningmedia.org/resource/ean08.sci.life.eco.athabaskan/a-subsistence-culture-impacted-by-climate-change/.

28. Armstrong, "Climate Change Communication."

29. Armstrong.

30. Lorraine Whitmarsh, "What's in a Name? Commonalities and Differences in Public Understanding of 'Climate Change' and 'Global Warming,'" *Public Understanding of Science*, 2008, http://pus.sagepub.com/content/early/2008/09/16/0963662506073088.short.

31. Jonathon P. Schuldt, Sara H. Konrath, and Norbert Schwarz, "'Global Warming' or 'Climate Change'? Whether the Planet Is Warming Depends on Question Wording," *Public Opinion Quarterly* 75, no. 1 (2011): 115–24.

32. Schuldt and Roh, "Media Frames and Cognitive Accessibility."

33. Jonathon P. Schuldt, Sungjong Roh, and Norbert Schwarz, "Questionnaire Design Effects in Climate Change Surveys: Implications for the Partisan Divide," *ANNALS of the American Academy of Political and Social Science* 658, no. 1 (March 1, 2015): 67–85, doi:10.1177/0002716214555066.

34. Ezra M. Markowitz and Azim F. Shariff, "Climate Change and Moral Judgement," *Nature Climate Change* 2, no. 4 (2012): 243–47.

35. Edward W. Maibach et al., "The Francis Effect: How Pope Francis Changes the Conversation about Global Warming," George Mason University Center for Climate Change Communication: George Mason University and Yale University, November 2015, http://environment.yale.edu/climate-communication/files/The_Francis_Effect.pdf.

36. Jonathon P. Schuldt et al., "Brief Exposure to Pope Francis Heightens Moral Beliefs about Climate Change," *Climatic Change* 141, no. 2 (March 1, 2017): 167–77, https://doi.org/10.1007/s10584-016-1893-9.

37. Asheley R. Landrum et al., "Processing the Papal Encyclical through Perceptual Filters: Pope Francis, Identity-Protective Cognition, and Climate Change Concern," *Cognition* 166, Supplement C (September 1, 2017): 1–12, https://doi.org/10.1016/j.cognition.2017.05.015.

38. William Adlong and Elaine Dietsch, "Environmental Education and the Health Professions: Framing Climate Change as a Health Issue," *Environmental Education Research* 21, no. 5 (2015): 687–709.

39. Armstrong, "Climate Change Communication."

40. Gill Ereaut and Nat Segnit, *Warm Words: How We Are Telling the Climate Story and Can We Tell It Better?*, Institute for Public Policy Research, London, 2006.

41. Thomas Dietz, Paul C. Stern, and Elke U. Weber, "Reducing Carbon-Based Energy Consumption through Changes in Household Behavior," *Daedalus* 142, no. 1 (2013): 78–89.

42. Dietz, Stern, and Weber.

43. Janis L. Dickinson, Rhiannon Crain, Steve Yalowitz, and Tammy M. Cherry, "How Framing Climate Change Influences Citizen Scientists' Intentions to Do Something About It," *Journal of Environmental Education* 44, no. 3 (January 2013): 145–58, doi:10.1080/00958964.2012.742032.

44. Dietz, Stern, and Weber, "Reducing Carbon-Based Energy Consumption."

45. Armstrong, "Climate Change Communication."

46. Teresa A. Myers, Matthew C. Nisbet, Edward W. Maibach, and Anthony A. Leiserowitz, "A Public Health Frame Arouses Hopeful Emotions about Climate Change: A Letter," *Climatic Change* 113 (2012): 1105–12, doi:10.1007/s10584-012-0513-6.

47. Louise Chawla and Debra Flanders Cushing, "Education for Strategic Environmental Behavior," *Environmental Education Research* 13, no. 4 (2007): 444; Albert Bandura, *Self-Efficacy: The Exercise of Control*, New York: W. H. Freeman and Company, 1997.

48. Morton et al., "Future That May (or May Not) Come."

49. C. R. Snyder et al., "The Will and the Ways: Development and Validation of an Individual-Differences Measure of Hope," *Journal of Personality and Social Psychology* 60, no. 4 (1991): 570–85.

50. Maria Ojala, "Hope and Climate Change: The Importance of Hope for Environmental Engagement among Young People," *Environmental Education Research* 18, no. 5 (2012): 625–42, doi:10.1080/13504622.2011.637157.

51. Ojala; Christine Li and Martha C. Monroe, "Development and Validation of the Climate Change Hope Scale for High School Students," *Environment and Behavior*, May 16, 2017, 0013916517708325, https://doi.org/10.1177/0013916517708325.

52. Stevenson and Peterson, "Motivating Action through Fostering Climate Change Hope and Concern and Avoiding Despair among Adolescents," *Sustainability* 8, no. 1 (January 2016): 6, https://doi.org/10.3390/su8010006v.

53. Stevenson and Peterson.

54. Myers et al., "Public Health Frame."

55. Adlong and Dietsch, "Environmental Education and the Health Professions"; Jonathon P. Schuldt, Katherine A. McComas, and Sahara E. Byrne, "Communicating about Ocean Health: Theoretical and Practical Considerations," *Philosophical Transactions of the Royal Society B* 371, no. 1689 (2016): 20150214.

56. *Project WET: Curriculum and Activity Guide 2.0* (Bozeman, MT: Project WET Foundation, 2011).

57. Adam Corner, Ezra Markowitz, and Nick Pidgeon, "Public Engagement with Climate Change: The Role of Human Values," *Wiley Interdisciplinary Reviews: Climate Change* 5, no. 3 (May 1, 2014): 411–22, doi:10.1002/wcc.269.

58. Jesse Graham, Jonathan Haidt, and Brian A. Nosek, "Liberals and Conservatives Rely on Different Sets of Moral Foundations," *Journal of Personality and Social Psychology* 96, no. 5 (2009): 1029.

59. Shalom Schwartz, "Are There Universal Aspects in the Structure and Contents of Human Values?," *Journal of Social Issues* 50, no. 4 (1994): 19–45.

60. Judith I. M. de Groot and Linda Steg, "Relationships between Value Orientations, Self-Determined Motivational Types and Pro-environmental Behaviors," *Journal of Environmental Psychology* 30 (2010): 368–78.

61. Corner, Markowitz, and Pidgeon, "Public Engagement with Climate Change."

62. Rachel Howell, "It's Not (Just) 'the Environment, Stupid!' Values, Motivations, and Routes to Engagement of People Adopting Lower-Carbon Lifestyles," *Global Environmen-*

tal Change 23 (2013): 281–90; Rachel Howell and Simon Allen, "People and Planet: Values, Motivations and Formative Influences of Individuals Acting to Mitigate Climate Change," *Environmental Values* 26, no. 2 (April 1, 2017): 131–55, https://doi.org/10.3197/096327117X14847335385436.

63. Thomas Dietz, Amy Dan, and Rachael Shwom, "Support for Climate Change Policy: Social Psychological and Social Structural Influences," *Rural Sociology* 72, no. 2 (2007): 185–214.

64. Raymond De Young, "New Ways to Promote Pro-environmental Behavior: Expanding and Evaluating Motives for Environmentally Responsible Behavior," *Journal of Social Issues* 56, no. 3 (2000): 509–26.

65. Shelby A. Krantz and Martha C. Monroe, "Message Framing Matters: Communicating Climate Change with Forest Landowners," *Journal of Forestry* 114, no. 2 (March 2016): 108–15, doi:http://dx.doi.org.proxy.library.cornell.edu/10.5849/jof.14-057.

66. Krantz and Monroe.

67. Aneeta Rattan, K. Savani, and Rainer Romero-Canyas, "Motivating Environmental Behavior by Framing Carbon Offset Requests Using Culturally Relevant Frames," Association of Psychological Science, New York, May 2015.

68. Robert D. Benford and David A. Snow, "Framing Processes and Social Movements: An Overview and Assessment," *Annual Review of Sociology* 26, no. 2000 (2010): 611–39.

9. USING METAPHOR AND ANALOGY IN CLIMATE CHANGE COMMUNICATION

1. "Metaphor—Examples and Definition of Metaphor," *Literary Devices*, June 14, 2013, http://literarydevices.net/metaphor/.

2. George Lakoff and Mark Johnson, *Metaphors We Live by* (Chicago: University of Chicago Press, 1980).

3. Xiang Chen, "The Greenhouse Metaphor and the Greenhouse Effect: A Case Study of a Flawed Analogous Model," in *Philosophy and Cognitive Science*, ed. Lorenzo Magnani and Ping Li (Berlin: Springer-Verlag, 2012), 105–14.

4. Chris Russill, "Temporal Metaphor in Abrupt Climate Change Communication: An Initial Effort at Clarification," in *The Economic, Social, and Political Elements of Climate Change*, ed. Walter Leal Filho (Berlin: Springer-Verlag, 2011), 16; Andrew Volmert, "Getting to the Heart of the Matter: Using Metaphorical and Causal Explanation to Increase Public Understanding of Climate and Ocean Change," FrameWorks Institute, May 2014, http://www.frameworksinstitute.org/assets/files/PDF_oceansclimate/occ_metaphor_report.pdf.

5. Volmert, "Getting to the Heart of the Matter."

6. Chen, "Greenhouse Metaphor and the Greenhouse Effect."

7. David Archer and Victor Brovkin, "The Millennial Atmospheric Lifetime of Anthropogenic CO_2," *Climatic Change* 90, no. 3 (October 1, 2008): 283–97, https://doi.org/10.1007/s10584-008-9413-1.

8. Archer and Brovkin.

9. Reinders Duit, "On the Role of Analogies and Metaphors in Learning Science," *Science Education* 75, no. 6 (November 1, 1991): 649–72, https://doi.org/10.1002/sce.3730750606.

10. Peter J. Aubusson, Allan G. Harrison, and Stephen M. Ritchie, "Metaphor and Analogy: Serious Thought in Science Education," in *Metaphor and Analogy in Science Education*, ed. Peter J. Aubusson, Allan G. Harrison, and Peter J. Aubusson (Dordrecht, Netherlands: Springer, 2006), 1–9.

11. Robert Kunzig, "The Carbon Bathtub," National Geographic, 2009, http://ngm.nationalgeographic.com/big-idea/05/carbon-bath.

12. "Steroids, Baseball and Climate Change," CLEAN: Climate Literacy & Energy Awareness Network, accessed June 17, 2016, http://cleanet.org/resources/43778.html.

13. Climate Interactive, "Carbon Bathtub," accessed December 31, 2017, https://www.climateinteractive.org/wp-content/uploads/2014/03/bathtub_CO2.jpg.

14. Nita A. Paris and Shawn M. Glynn, "Elaborate Analogies in Science Text: Tools for Enhancing Preservice Teachers' Knowledge and Attitudes," *Contemporary Educational Psychology* 29, no. 3 (July 2004): 230–47, doi:10.1016/S0361-476X(03)00033-X.

15. Paris and Glynn, 242.

16. Shawn M. Glynn, "Making Science Concepts Meaningful to Students: Teaching with Analogies," in *Four Decades of Research in Science Education: From Curriculum Development to Quality Improvement*, ed. Silke Mikelskis-Seifert and Ute Ringelband (Münster: Waxmann Verlag, 2008), 113.

17. Shawn M. Glynn, "The Teaching-with-Analogies Model," NSTA WebNews Digest, Science and Children: Methods and Strategies, http://www.nsta.org/publications/news/story.aspx?id=53640.

18. Glynn, "Making Science Concepts Meaningful to Students."

19. Volmert, "Getting to the Heart of the Matter."

20. Glynn, "Making Science Concepts Meaningful to Students."

10. CLIMATE CHANGE MESSENGERS

1. Ariel Malka, Jon A. Krosnick, and Gary Langer, "The Association of Knowledge with Concern about Global Warming: Trusted Information Sources Shape Public Thinking," *Risk Analysis* 29, no. 5 (May 2009): 633–47; Andrew J. Hoffman, *How Culture Shapes the Climate Change Debate* (Stanford, CA: Stanford University Press, 2015).

2. Marc J. Stern and Kimberly J. Coleman, "The Multidimensionality of Trust: Applications in Collaborative Natural Resource Management," *Society and Natural Resources* 28 (2015): 117–32.

3. Jean Goodwin and Michael F. Dahlstrom, "Communication Strategies for Earning Trust in Climate Change Debates," *Wiley Interdisciplinary Reviews: Climate Change* 5, no. 1 (January/February 2014), doi: 10.1002/wcc.262.

4. Goodwin and Dahlstrom.

5. Susanne C. Moser, "Communicating Climate Change: History, Challenges, Process and Future Directions," *Wiley Interdisciplinary Reviews: Climate Change* 1, no. 1 (January 1, 2010): 31–53, doi:10.1002/wcc.11.

6. Matthew C. Nisbet, "Public Opinion and Participation," in *The Oxford Handbook of Climate Change and Society*, ed. John S. Dryzek, Richard B. Noorgaard, and David Schlossberg (Oxford: Oxford University Press, 2011), 357.

7. Shelby A. Krantz and Martha C. Monroe, "Message Framing Matters: Communicating Climate Change with Forest Landowners," *Journal of Forestry* 114, no. 2 (March 2016): 108–15, http://dx.doi.org.proxy.library.cornell.edu/10.5849/jof.14-057.

8. Martha C. Monroe, Richard R. Plate, Damien C. Adams, and Deborah J. Wojcik, "Harnessing Homophily to Improve Climate Change Education," *Environmental Education Research* 21, no. 2 (2015): 221–38, doi:10.1080/13504622.2014.910497.

9. Edward W. Maibach et al., "Identifying Like-Minded Audiences for Global Warming Public Engagement Campaigns: An Audience Segmentation Analysis and Tool Development," *PLOS ONE* 6, no. 3 (2011): e17571, https://doi.org/10.1371/journal.pone.0017571.

10. John Fraser, Anthony Taylor, Erin Johnson, and Jessica Sickler, "The Relative Credibility of Zoo-Affiliated Spokespeople for Delivering Conservation Messages," *Curator: The Museum Journal* 51, no. 4 (October 1, 2008): 407–18, https://doi.org/10.1111/j.2151-6952.2008.tb00326.x.

11. CLIMATE CHANGE EDUCATION AT THE MARINE MAMMAL CENTER, SAUSALITO, CALIFORNIA

1. Jerry Luebke, Susan Clayton, Carol Saunders, Jennifer Matiasek, Lisa-Anne Kelly, and Alejandro Grajal, *Global Climate Change as Seen by Zoo and Aquarium Visitors* (Brookfield, IL: Chicago Zoological Society, 2012), http://citeseerx.ist.psu.edu/viewdoc/download?doi=10.1.1.460.1111&rep=rep1&type=pdf.

2. Susan Clayton, Jerry Luebke, Carol Saunders, Jennifer Matiasek, and Alejandro Grajal, "Connecting to Nature at the Zoo: Implications for Responding to Climate Change," *Environmental Education Research* 20, no. 4 (July 4, 2014): 460–75, doi:10.1080/13504622.2013.816267.

12. CLIMATE CHANGE LITERACY, ACTION, AND POSITIVE YOUTH DEVELOPMENT IN KENTUCKY

1. Katherine M. Emmons, "Perspectives on Environmental Action: Reflection and Revision through Practical Experience," *Journal of Environmental Education* 29, no. 1 (November 1, 1997): 34–44, doi:10.1080/00958969709599105.

2. Tania M. Schusler and Marianne E. Krasny, "Environmental Action as Context for Youth Development," *Journal of Environmental Education* 41, no. 4 (January 1, 2010): 208–23.

3. Scott Wartman, "McConnell: Don't Expect Much from Congress This Year," Cincinnati.com, March 7, 2014, http://www.cincinnati.com/story/news/politics/elections/2014/03/07/mcconnell-expect-much-congress-year/6170921/.

13. BUILDING SOIL TO CAPTURE CARBON IN A SCHOOL GARDEN IN NEW MEXICO

1. Eric Plutzer, Mark McCaffrey, A. Lee Hannah, Joshua Rosenau, Minda Berbeco, and Ann H. Reid, "Climate Confusion among U.S. Teachers," *Science*, February 12, 2016, doi:10.1126/science.aab3907.

2. Martha Monroe, Annie Oxarart, and Richard Plate, "A Role for Environmental Education in Climate Change for Secondary Science Educators," *Applied Environmental Education and Communication* 12, no. 1 (2013), doi:10.1080/1533015X.2013.795827.

3. Gary Paul Nabhan, *Growing Food in a Hotter, Drier Land: Lessons from Desert Farmers on Adapting to Climate Uncertainty* (White River Junction, VT: Chelsea Green, 2013).

14. PSYCHOLOGICAL RESILIENCE IN DENVER, COLORADO

1. Mary Pipher, *The Green Boat: Reviving Ourselves in Our Capsized Culture* (New York: Riverhead Books, 2013).

Select Bibliography

Albert, Simon, Alistair Grinham, Badin Gibbes, Javier Leon, and John Church. "Sea Level Rise Swallows 5 Whole Pacific Islands." *Scientific American*, Conversation, May 9, 2016. https://www.scientificamerican.com/article/sea-level-rise-swallows-5-whole-pacific-islands/.

Andrews, Travis M. "Trump Calls Mayor of Shrinking Chesapeake Island and Tells Him Not to Worry about It." *Washington Post*, June 14, 2017. https://www.washingtonpost.com/news/morning-mix/wp/2017/06/14/trump-calls-mayor-of-shrinking-chesapeake-island-and-tells-him-not-to-worry-about-it/?utm_term=.bfcc3c6e0df8.

Archer, David, and Victor Brovkin. "The Millennial Atmospheric Lifetime of Anthropogenic CO_2." *Climatic Change* 90, no. 3 (October 1, 2008): 283–97. doi:10.1007/s10584-008-9413-1.

Armstrong, Anne Katherine. "Climate Change Communication and Environmental Education: From Research to Practice." Master's thesis, Cornell University, 2017. https://search.proquest.com/pqdtlocal1006599/docview/1985045643/D6FE2667A3B04C16PQ/2.

Aubusson, Peter J., Allan G. Harrison, and Stephen M. Ritchie. "Metaphor and Analogy: Serious Thought in Science Education." In *Metaphor and Analogy in Science Education*, edited by Peter J. Aubusson, Allan G. Harrison, and Peter J. Aubusson, 1–9. Dordrecht, Netherlands: Springer, 2006.

Bandura, Albert. *Self-Efficacy: The Exercise of Control*. New York: W. H. Freeman and Company, 1997.

Bandura, Albert. *Social Learning Theory*. Englewood Cliffs, NJ: Prentice-Hall, 1977.

Benford, Robert D., and David A. Snow. "Framing Processes and Social Movements: An Overview and Assessment." *Annual Review of Sociology* 26, no. 2000 (2010): 611–39.

Berwyn, Bob. "Why Is Antarctica's Sea Ice Growing While the Arctic Melts? Scientists Have an Answer." Inside Climate News, May 31, 2016.

Bliuc, Ana-Maria, Craig McGarty, Emma F. Thomas, Girish Lala, Mariette Berndsen, and RoseAnne Misajon. "Public Division about Climate Change Rooted in Conflicting Socio-Political Identities." *Nature Climate Change* 5, no. 3 (March 2015): 226–29. doi:10.1038/nclimate2507.

Brügger, Adrian, Thomas A. Morton, and Suraje Dessai. "'Proximising' Climate Change Reconsidered: A Construal Level Theory Perspective." *Journal of Environmental Psychology* 46, Supplement C (June 1, 2016): 125–42. doi:10.1016/j.jenvp.2016.04.004.

Buntgen, Ulf, Vladimir S. Myglan, Fredrik Charpentier Ljungqvist, Michael McCormick, Nicola Di Cosmo, Michael Sigl, Johann Jungclaus, et al. "Cooling and Societal Change during the Late Antique Little Ice Age from 536 to around 660 AD." *Nature Geoscience* 9, no. 3 (2016): 231–36. doi:10.1038/ngeo2652.

Busch, K. C. "Polar Bears or People? Exploring Ways in Which Teachers Frame Climate Change in the Classroom." *International Journal of Science Education* 6, no. 2 (April 2, 2016): 137–65. doi:10.1080/21548455.2015.1027320.

Cave, Damien, and Justin Gillis. "Large Sections of Australia's Great Reef Are Now Dead, Scientists Find." *New York Times*, March 15, 2017. https://www.nytimes.com/2017/03/15/science/great-barrier-reef-coral-climate-change-dieoff.html.

Centers for Disease Control and Prevention. "Climate Change Increases the Number and Geographic Range of Disease-Carrying Insects and Ticks." N.d. https://www.cdc.gov/climateandhealth/pubs/vector-borne-disease-final_508.pdf.

Chan, Kenny K., and Shekhar Misra. "Characteristics of the Opinion Leader: A New Dimension." *Journal of Advertising* 19, no. 3 (1990): 53–60.

Chawla, Louise, and Debra Flanders Cushing. "Education for Strategic Environmental Behavior." *Environmental Education Research* 13, no. 4 (September 1, 2007): 437–52. doi:10.1080/13504620701581539.

Children and Nature Network. 2018. http://www.childrenandnature.org/.

China Daily. "Green Platform Turns Virtual Trees into Desert Guardians." March 29, 2017. http://www.chinadaily.com.cn/regional/2017-03/29/content_28720126.htm.

Climate and Urban Systems Partnership. "Pittsburgh." Accessed January 1, 2018. http://www.cuspproject.org/cities/pittsburgh#.WkpdK9-nHIU.

Climate Interactive. "Carbon Bathtub." Accessed December 31, 2017. https://www.climateinteractive.org/wp-content/uploads/2014/03/bathtub_CO2.jpg.

Community and Regional Resilience Institute (CARRI). *Definitions of Community Resilience: An Analysis*. Washington, DC: Community and Regional Resilience Institute, 2013.

Community Climate Change Fellows. *Community Climate Change Education: A Mosaic of Approaches*. Edited by Marna Hauk and Elizabeth Pickett. Washington, DC: North American Association for Environmental Education, 2016. https://naaee.org/eepro/resources/community-climate-change-education.

Cornell Cooperative Extension. "Energy Smart Community Tompkins." 2018. http://ccetompkins.org/energy/energy-smart-community-tompkins/faqs.

Dahlman, LuAnn. "Climate Change: Ocean Heat Content." NOAA, Climate.gov. 2015. https://www.climate.gov/news-features/understanding-climate/climate-change-ocean-heat-content.

De Young, Raymond. "New Ways to Promote Pro-environmental Behavior: Expanding and Evaluating Motives for Environmentally Responsible Behavior." *Journal of Social Issues* 56, no. 3 (2000): 509–26.

Dietz, Thomas, Paul C. Stern, and Elke U. Weber. "Reducing Carbon-Based Energy Consumption through Changes in Household Behavior." *Daedalus* 142, no. 1 (2013): 78–89.

Downey, Laura, Susan Jane Gentile, Karen S. Hollweg, Jennifer Hubbard-Sanchez, Christopher Johnson, L. Kumler, Lisa LaRoque, Kristen Poppleton, Deborah Shiflett-Fitton, and Jay Shuttleworth, eds. "Advancing Climate Change Environmental Education: Resources and Suggestions." EE Capacity, Cornell University, Civic Ecology Lab, North American Association of Environmental Education, 2013. http://www.eecapacity.net.

Dubois, Bryce, and Marianne E. Krasny. "Educating with Resilience in Mind: Addressing Climate Change in Post-Sandy New York City." *Journal of Environmental Education* 47, no. 4 (April 29, 2016): 1–16. doi:10.1080/00958964.2016.1167004.

Dunlap, Riley E., Aaron M. McCright, and Jerrod H. Yarosh. "The Political Divide on Climate Change: Partisan Polarization Widens in the U.S." *Environment* 58, no. 5 (September 2, 2016): 4–23. doi:10.1080/00139157.2016.1208995.

Egan, Patrick J., and Megan Mullin. "Climate Change: U.S. Public Opinion." *Annual Review of Political Science* 20 (2017): 209–27.

Eizenberg, Efrat. *From the Ground Up: Community Gardens in New York City and the Politics of Spatial Transformation*. New York: Routledge, 2016.

Ellsworth, Phoebe, and Klaus R. Scherer. "Appraisal Processes in Emotion." In *Handbook of Affective Sciences*, edited by Richard J. Davidson, Klaus R. Scherer, and H. Hill Goldsmith, 572–95. Oxford: Oxford University Press, 2003.

Energy Star. "Energy Savings at Home." N.d. https://www.energystar.gov/index.cfm?c=heat_cool.pr_hvac.

EPA (Environmental Protection Agency). "Climate Change Indicators: Arctic Sea Ice." 2016. https://www.epa.gov/climate-indicators/climate-change-indicators-arctic-sea-ice.

——. "Greenhouse Gas Emissions." 2017. https://www.epa.gov/ghgemissions/overview-greenhouse-gases.

——. "Sources of Greenhouse Gas Emissions." 2017. https://www.epa.gov/ghgemissions/sources-greenhouse-gas-emissions.

Fernbach, Philip M., Todd Rogers, Craig R. Fox, and Steven A. Sloman. "Political Extremism Is Supported by an Illusion of Understanding." *Psychological Science* 24, no. 6 (June 1, 2013): 939–46. doi:10.1177/0956797612464058.

Festinger, Leon. "Cognitive Dissonance." *Scientific American* 207 (1962): 93–102. doi:10.1037/10318-001.

Foote, Eunice. "Circumstances Affecting the Heat of the Sun's Rays." *American Journal of Science and the Arts* 22 (1856): 382–83.

Gallup. "Worry about Terror Attacks in U.S. High, but Not Top Concern." Gallup.com, March 23, 2016. http://news.gallup.com/poll/190253/worry-terror-attacks-high-not-top-concern.aspx.

Gifford, Robert, and Reuven Sussman. "Environmental Attitudes." In *The Oxford Handbook of Environmental and Conservation Psychology*. Oxford Handbooks Online, September 28, 2012. doi:10.1093/oxfordhb/9780199733026.013.0004.

Glynn, Shawn M. "Making Science Concepts Meaningful to Students: Teaching with Analogies." In *Four Decades of Research in Science Education: From Curriculum Development to Quality Improvement*, edited by Silke Mikelskis-Seifert and Ute Ringelband, 113–26. Münster: Waxmann Verlag, 2008.

Goodwin, Jean, and Michael F. Dahlstrom. "Communication Strategies for Earning Trust in Climate Change Debates." *Wiley Interdisciplinary Reviews: Climate Change* 5, no. 1 January/February 2014. doi:10.1002/wcc.262.

Graham, Steve. "John Tyndall (1820–1893)." NASA Earth Observatory, October 8, 1999. https://earthobservatory.nasa.gov/Features/Tyndall/.

Green, Carie, Darius Kalvaitis, and Anneliese Worster. "Recontextualizing Psychosocial Development in Young Children: A Model of Environmental Identity Development." *Environmental Education Research*, August 14, 2015, 1–24. doi:10.1080/13504622.2015.1072136.

Ham, Sam H. *Environmental Interpretation: A Practical Guide for People with Big Ideas and Small Budgets*. Golden, CO: North American, 1992.

Harlan, Sharon L., Juan H. Declet-Barreto, William L. Stefanov, and Diana B. Petitti. "Neighborhood Effects on Heat Deaths: Social and Environmental Predictors of Vulnerability in Maricopa County, Arizona." *Environmental Health Perspectives* 121 (2013): 197–204. doi:10.1289/ehp.1104625.

Harmon, Amy. "Climate Science Meets a Stubborn Obstacle: Students." *New York Times*, June 4, 2017. https://www.nytimes.com/2017/06/04/us/education-climate-change-science-class-students.html.

Harpp, Karen. "How Do Volcanoes Affect World Climate?" *Scientific American* online, 2002. https://www.scientificamerican.com/article/how-do-volcanoes-affect-w/.

Hart, P. Sol, and Lauren Feldman. "Threat without Efficacy? Climate Change on U.S. Network News." *Science Communication* 36, no. 3 (February 2014): 325–51. doi:10.1177/1075547013520239.

Hart, P. Sol, and Erik C. Nisbet. "Boomerang Effects in Science Communication: How Motivated Reasoning and Identity Cues Amplify Opinion Polarization about Climate Mitigation Policies." *Communication Research* 39, no. 6 (December 1, 2012): 701–23. doi:10.1177/0093650211416646.

Hartmann, Dennis L., Albert M. G. Klein Tank, Matilde Rusticucci, Lisa V. Alexander, Stefan Brönnimann, Yassine Abdul-Charabi, Frank J. Dentener, et al. "Observations: Atmosphere and Surface." In *Climate Change 2013: The Physical Science Basis; Contribution of Working Group I to the Fifth Assessment Report of the Intergovernmental Panel on Climate Change*, edited by T. F. Stocker, D. Qin, G.-K. Plattner, M. Tignor, S. K. Allen, J. Boschung, A. Nauels, Y. Xia, V. Bex, and P. M. Midgley. Cambridge: Cambridge University Press.

Heat Smart Tompkins. "Harnessing the Efficiencies of Home Energy Performance." 2017. http://www.solartompkins.org/home-energy-and-heat-pump-faqs.html.

Heberlein, Thomas A. *Navigating Environmental Attitudes*. New York: Oxford University Press, 2012.

Heberlein, Thomas A., and J. Stanley Black. "Attitudinal Specificity and the Prediction of Behavior in a Field Setting." *Journal of Personality and Social Psychology* 33, no. 4 (1976): 474.

Hoffman, Andrew J. *How Culture Shapes the Climate Change Debate*. Stanford, CA: Stanford University Press, 2015.

Holling, C. S. "Resilience and Stability of Ecological Systems." *Annual Review of Ecology and Systematics* 4 (1973): 1–23.

Holpuch, Amanda. "Alaskan Village Threatened by Rising Sea Levels Votes for Costly Relocation." *Guardian*, August 18, 2016. https://www.theguardian.com/us-news/2016/aug/18/alaska-shishmaref-vote-move-coastal-erosion-rising-sea-levels.

Hönisch, Bärbel, Andy Ridgwell, Daniela N. Schmidt, Ellen Thomas, Samantha J. Gibbs, Andy Sluijs, Richard Zeebe, et al. "The Geological Record of Ocean Acidification." *Science*, 335, no. 6072 (March 2012): 1058–63. doi:10.1126/science.1208277.

Hornsey, Matthew J., Emily A. Harris, Paul G. Bain, and Kelly S. Fielding. "Meta-Analyses of the Determinants and Outcomes of Belief in Climate Change." *Nature Climate Change* 6 (2016): 622–26. http://www.nature.com/nclimate/journal/vaop/ncurrent/full/nclimate2943.html?WT.feed_name=subjects_attribution.

Howe, Peter, Matto Mildenburger, Jennifer R. Marlon, and Anthony Leiserowitz. "Geographic Variation in Opinions on Climate Change at State and Local Scales in the USA." *Nature Climate Change* 5, no. 6 (June 2015): 596–603.

Howell, Rachel, and Simon Allen. "People and Planet: Values, Motivations and Formative Influences of Individuals Acting to Mitigate Climate Change." *Environmental Values* 26, no. 2 (April 1, 2017): 131–55. doi:10.3197/096327117X14847335385436.

"HS-ESS3 Earth and Human Activity." Next Generation Science Standards. Accessed March 7, 2016. http://www.nextgenscience.org/dci-arrangement/hs-ess3-earth-and-human-activity.

Hu, Winnie. "No Room on a Bike Rack? Not a Problem for These Smart Bikes." *New York Times*, August 28, 2017. https://www.nytimes.com/2017/08/28/nyregion/bike-sharing-spin-gps-dockless-citi-bike.html?mcubz=0&_r=0.

Hungerford, Harold R., and Trudi L. Volk. "Changing Learner Behavior through Environmental Education." *Journal of Environmental Education* 21, no. 3 (March 1, 1990): 8–21. doi:10.1080/00958964.1990.10753743.

Intergovernmental Panel on Climate Change (IPCC). *Climate Change 2014: Impacts, Adaptation, and Vulnerability. Part A: Global and Sectoral Aspects. Contribution of Working Group II to the Fifth Assessment Report of the Intergovernmental Panel on Climate Change.* Edited by Christopher B. Field, Vicente R. Barros, David Jon Dokken, Katharine J. Mach, Michael D. Mastrandrea, T. Eren Bilir, Monalisa Chatterjee, et al. Cambridge: Cambridge University Press, 2014. https://www.ipcc.ch/report/ar5/.

——. *Climate Change 2014: Synthesis Report; Summary for Policymakers.* Geneva, Switzerland, 2014. https://www.ipcc.ch/pdf/assessment-report/ar5/syr/AR5_SYR_FINAL_SPM.pdf.

Jacobson, Susan K. *Communication Skills for Conservation Professionals.* 2nd ed. Washington, DC: Island Press, 2009.

Johnson-Pynn, Julie S., and Laura R. Johnson. "Successes and Challenges in East African Conservation Education." *Journal of Environmental Education* 36, no. 2 (2005): 25–39.

Jones, Nicola. "How the World Passed a Carbon Threshold and Why It Matters." *Yale Environment 360*, January 26, 2017. https://e360.yale.edu/features/how-the-world-passed-a-carbon-threshold-400ppm-and-why-it-matters.

Kelly, Anne E., and Michael L. Goulden. 2008. "Rapid Shifts in Plant Distribution with Recent Climate Change." *PNAS* 105 (33): 11823–26.

Kentucky Department of Education. "Next Generation Science Standards (NGSS)." Accessed December 30, 2017. https://education.ky.gov/curriculum/conpro/science/Pages/Next-Generation-Science-Standards.aspx.

Krantz, Shelby A., and Martha C. Monroe. "Message Framing Matters: Communicating Climate Change with Forest Landowners." *Journal of Forestry* 114, no. 2 (March 2016): 108–15. http://dx.doi.org.proxy.library.cornell.edu/10.5849/jof.14-057.

Krasny, Marianne E., and Bryce DuBois. "Climate Adaptation Education: Embracing Reality or Abandoning Environmental Values." *Environmental Education Research*, June 16, 2016, 1–12. doi:10.1080/13504622.2016.1196345.

Krasny, Marianne E., Leigh Kalbacker, Richard C. Stedman, and Alex Russ. "Measuring Social Capital among Youth: Applications in Environmental Education." *Environmental Education Research* 21, no. 1 (2015): 1–23.

Krasny, Marianne E., John Carey, Bryce DuBois, Caroline Lewis, John Fraser, Kari Fulton, Billy Spitzer, et al. *Climate Change and Environmental Education: Framing Perspectives.* Cornell University Civic Ecology Lab report, February 1, 2015.

Kunkle, Kristen, and Martha Monroe. "Misconceptions and Psychological Mechanisms of Climate Change Communication." In *Across the Spectrum: Resources for Environmental Educators*, edited by Martha Monroe and Marianne Krasny, 3rd ed. North American Association of Environmental Education, 2016.

Landrum, Asheley R., Robert B. Lull, Heather Akin, Ariel Hasell, and Kathleen Hall Jamieson. "Processing the Papal Encyclical through Perceptual Filters: Pope Francis, Identity-Protective Cognition, and Climate Change Concern." *Cognition* 166, no. Supplement C (September 1, 2017): 1–12. doi:10.1016/j.cognition.2017.05.015.

Latimer, Christopher, and Karen Hempson. "Using Deliberation in the Classroom: A Teaching Pedagogy to Enhance Student Knowledge, Opinion Formation, and Civic Engagement." *Journal of Political Science Education* 8 (2012): 372–88.

Lee, Jeffrey. "Milankovitch Cycles." Encyclopedia of Earth, July 7, 2010. http://www.eoearth.org/view/article/154612/.

Leiserowitz, Anthony, Edward Maibach, Connie Roser-Renouf, Seth Rosenthal, Matthew Cutler, and John Kotcher. *Climate Change in the American Mind: October 2017*. New Haven, CT: Yale Project on Climate Change Communication, Yale University and George Mason University, 2017. http://climatecommunication.yale.edu/publications/climate-change-american-mind-october-2017/.

Levin, Irwin P., and Gary J. Gaeth. "How Consumers Are Affected by the Framing of Attribute Information before and after Consuming the Product." *Journal of Consumer Research* 15, no. 3 (December 1, 1988): 374–78. doi:10.1086/209174.

Li, Christine, and Martha C. Monroe. "Development and Validation of the Climate Change Hope Scale for High School Students." *Environment and Behavior*, May 16, 2017. doi:10.1177/0013916517708325.

Linden, Sander L. van der, Anthony A. Leiserowitz, Geoffrey D. Feinberg, and Edward W. Maibach. "How to Communicate the Scientific Consensus on Climate Change: Plain Facts, Pie Charts or Metaphors?" *Climatic Change* 126, nos. 1–2 (September 1, 2014): 255–62. doi:10.1007/s10584-014-1190-4.

——. "The Scientific Consensus on Climate Change as a Gateway Belief: Experimental Evidence." *PLOS ONE* 10, no. 2 (February 25, 2015): e0118489. doi:10.1371/journal.pone.0118489.

Lindsey, Rebecca. "Global Impacts of El Niño and La Niña." NOAA, Climate.gov, February 9, 2016. https://www.climate.gov/news-features/featured-images/global-impacts-el-ni%C3%B1o-and-la-ni%C3%B1a.

Lorenzoni, Irene, Sophie Nicholson-Cole, and Lorraine Whitmarsh. "Barriers Perceived to Engaging with Climate Change among the UK Public and Their Policy Implications." *Global Environmental Change* 17, no. 3 (2007): 445–59.

Lowy, Benjamin. "When Rising Seas Transform Risk into Certainty." *New York Times*, April 18, 2017. https://www.nytimes.com/2017/04/18/magazine/when-rising-seas-transform-risk-into-certainty.html.

Lu, Hang, and Jonathon P. Schuldt. "Compassion for Climate Change Victims and Support for Mitigation Policy." *Journal of Environmental Psychology* 45 (2016): 192–200.

Maibach, Edward, Anthony Leiserowitz, Connie Roser-Renouf, and C. K. Mertz. "Identifying Like-Minded Audiences for Global Warming Public Engagement Campaigns: An Audience Segmentation Analysis and Tool Development." *PLOS ONE* 6, no. 3 (2011): e17571. doi:10.1371/journal.pone.0017571.

Maibach, Edward, Anthony Leiserowitz, Connie Roser-Renouf, Teresa Myers, Seth Rosenthal, and Geoff Feinberg. "The Francis Effect: How Pope Francis Changes the Conversation about Global Warming." Fairfax, VA: George Mason University Center for Climate Change Communication, George Mason University and Yale University, November 2015. http://environment.yale.edu/climate-communication/files/The_Francis_Effect.pdf.

Markowitz, Ezra M., and Azim F. Shariff. "Climate Change and Moral Judgement." *Nature Climate Change* 2, no. 4 (2012): 243–47.

Masten, Ann, and Jelena Obradović "Disaster Preparation and Recovery: Lessons from Research on Resilience in Human Development." *Ecology and Society* 13, no. 1 (2008).

McCright, Aaron M., and Riley E. Dunlap. "The Politicization of Climate Change and Polarization in the American Public's Views of Global Warming, 2001–2010." *Sociological Quarterly* 52, no. 2 (2011): 155–94.

Melillo, Jerry M., Terese Richmond, and Gary W. Yohe. *Climate Change Impacts in the United States: The Third National Climate Assessment.* U.S. Global Change Research Program, 2014. http://s3.amazonaws.com/nca2014/high/NCA3_Climate_Change_Impacts_in_the_United%20States_HighRes.pdf?download=1.

Modugno, Lindsay, Jeff Pace, and Dan Lidor. "The Effects of Climate Change and Sea Level Rise on the Coast." Sandy Hook Cooperative Research Programs, January 2015. https://marine.rutgers.edu/geomorph/geomorph/_pages/climatechange.html.

Monroe, Martha C., and Annie Oxarart, eds. *Southeastern Forests and Climate Change: A Project Learning Tree Secondary Environmental Education Module.* Gainesville: University of Florida and American Forest Foundation, 2014. http://sfrc.ufl.edu/extension/ee/climate/.

Monroe, Martha C., Richard R. Plate, Annie Oxarart, Alison Bowers, and Willandia A. Chaves. "Identifying Effective Climate Change Education Strategies: A Systematic Review of the Research." *Environmental Education Research*, 2017, 1–22.

Morton, Thomas A., Anna Rabinovich, Dan Marshall, and Pamela Bretschneider. "The Future That May (or May Not) Come: How Framing Changes Responses to Uncertainty in Climate Change Communications." *Global Environmental Change* 21, no. 1 (February 2011): 103–9. doi:10.1016/j.gloenvcha.2010.09.013.

Myers, Meghann. "Rising Oceans Threaten to Submerge 128 Military Bases: Report." *Navy Times*, July 29, 2016. https://www.navytimes.com/story/military/2016/07/29/rising-oceans-threaten-submerge-18-military-bases-report/87657780/.

Nabhan, Gary Paul. *Growing Food in a Hotter, Drier Land: Lessons from Desert Farmers on Adapting to Climate Uncertainty.* White River Junction, VT: Chelsea Green, 2013.

NASA (National Aeronautics and Space Administration). Climate. "Sea Ice Extent Sinks to Record Lows at Both Poles." August 6, 2017. https://www.nasa.gov/feature/goddard/2017/sea-ice-extent-sinks-to-record-lows-at-both-poles.

——. Global Climate Change: Vital Signs of the Planet. "Graphic: The Relentless Rise of Carbon Dioxide." Accessed January 23, 2018. https://climate.nasa.gov/climate_resources/24/.

——. Global Climate Change: Vital Signs of the Planet. "Responding to Climate Change." http://climate.nasa.gov/solutions/adaptation-mitigation.

NASA Earth Observatory. "Global Temperature Record Broken for Third Consecutive Year." 2017. https://earthobservatory.nasa.gov/IOTD/view.php?id=89469.

National Audubon Society. *Tools of Engagement: A Toolkit for Engaging People in Conservation.* Edited by Judy Braus. National Audubon Society, January 2011.

National Climate Assessment (NCA). "Ocean Acidification." 2014. http://nca2014.globalchange.gov/report/our-changing-climate/ocean-acidification.

National Network for Ocean and Climate Change Interpretation (NNOCCI). "The Problem with Solutions—and How to Fix It." 2017. http://climateinterpreter.org/features/problem-solutions%E2%80%94and-how-fix-it.

National Oceanic and Atmospheric Administration (NOAA). "Introduction to Storm Surge." N.d. https://www.nhc.noaa.gov/surge/surge_intro.pdf.

——. Shoreline Website. Glossary. "Living Shorelines." Accessed October 25, 2017. https://shoreline.noaa.gov/glossary.html.

——. "Storm Surge Overview." N.d. http://www.nhc.noaa.gov/surge/.

National Research Council (NRC). *Himalayan Glaciers: Climate Change, Water Resources, and Water Security.* Washington, DC: National Academies, 2012. https://www.nap.edu/read/13449/chapter/1.

New York State Department of Environmental Conservation. "A Guide to Local Action: Climate Smart Communities Certification." N.d. http://www.dec.ny.gov/energy/50845.html.

Nisbet, Matthew C. "Communicating Climate Change: Why Frames Matter for Public Engagement." *Environment: Science and Policy for Sustainable Development* 51, no. 2 (2009): 12–23.

North American Association for Environmental Education (NAAEE). "About EE and Why It Matters." May 19, 2015. https://naaee.org/about-us/about-ee-and-why-it-matters.

Ojala, Maria. "Hope and Climate Change: The Importance of Hope for Environmental Engagement among Young People." *Environmental Education Research* 18, no. 5 (2012): 625–42.

Otieno, Christine, Hans Spada, Katharina Liebler, Thomas Ludemann, Ulrich Deil, and Alexander Renkl. "Informing about Climate Change and Invasive Species: How the Presentation of Information Affects Perception of Risk, Emotions, and Learning." *Environmental Education Research* 20, no. 5 (September 3, 2014): 612–38. doi:10.1080/13504622.2013.833589.

Oyserman, Daphna. "Identity-Based Motivation: Implications for Action-Readiness, Procedural-Readiness, and Consumer Behavior." *Journal of Consumer Psychology* 19, no. 3 (July 2009): 250–60. doi:10.1016/j.jcps.2009.05.008.

Padhy, Susanta Kumar, Sidharth Sarkar, Mahima Panigrahi, and Surender Paul. "Mental Health Effects of Climate Change." *Indian Journal of Occupational and Environmental Medicine* 19, no. 1 (April 2015): 3–7. doi:10.4103/0019-5278.156997.

Paolisso, Michael, Ellen Douglas, Ashley Enrici, Paul Kirshen, Chris Watson, and Matthias Ruth. "Climate Change, Justice, and Adaptation among African American Communities in the Chesapeake Bay Region." *Weather, Climate, and Society* 4, no. 1 (January 1, 2012): 34–47. doi:10.1175/WCAS-D-11-00039.1.

Pearson, Adam R., Matthew T. Ballew, Sarah Naiman, and Jonathon P. Schuldt. "Race, Class, Gender and Climate Change Communication." *Oxford Research Encyclopedias: Climate Science*. April 26, 2017. http://climatescience.oxfordre.com/view/10.1093/acrefore/9780190228620.001.0001/acrefore-9780190228620-e-412.

Pearson, Adam R., Jonathon P. Schuldt, and Rainer Romero-Canyas. "Social Climate Science: A New Vista for Psychological Science." In *Perspectives on Psychological Science*. 2016. http://research.pomona.edu/sci/files/2016/06/PearsonSchuldtRomeroCanyas2016PPS-Social-Climate-Science.pdf.

Polar Science Center. "PIOMAS Arctic Sea Ice Volume Reanalysis." 2017. http://psc.apl.uw.edu/research/projects/arctic-sea-ice-volume-anomaly/.

Quest. "How Do Greenhouse Gases Work?" KQED Science, December 12, 2014. https://ww2.kqed.org/quest/2014/12/12/how-do-greenhouse-gases-work/.

Rattan, Aneeta, K. Savani, and Rainer Romero-Canyas. "Motivating Environmental Behavior by Framing Carbon Offset Requests Using Culturally Relevant Frames." Presented at the Association of Psychological Science, New York, May 2015.

Reuters. "Five Pacific Islands Lost to Rising Seas as Climate Change Hits." *Guardian*, May 10, 2016. https://www.theguardian.com/environment/2016/may/10/five-pacific-islands-lost-rising-seas-climate-change.

Rickard, Laura N., Z. Janet Yang, and Jonathon P. Schuldt. "Here and Now, There and Then: How 'Departure Dates' Influence Climate Change Engagement."

Global Environmental Change 38 (May 2016): 97–107. doi:10.1016/j.gloenvcha.2016.03.003.

Riebeek, Holli. "Global Warming." NASA Earth Observatory, June 3, 2010. http://earthobservatory.nasa.gov/Features/GlobalWarming/page4.php.

Rojas-Downing, M. Melissa, A. Pouyan Nejadhashemi, Timothy Harrigan, and Sean A. Woznicki. "Climate Change and Livestock: Impacts, Adaptation, and Mitigation." *Climate Risk Management* 16 (January 1, 2017): 145–63. doi:10.1016/j.crm.2017.02.001.

Rubenstein, Madeleine. "Emissions from the Cement Industry." *State of the Planet* (blog), May 9, 2012. Earth Institute, Columbia University. http://blogs.ei.columbia.edu/2012/05/09/emissions-from-the-cement-industry/.

Russill, Chris. "Temporal Metaphor in Abrupt Climate Change Communication: An Initial Effort at Clarification." In *The Economic, Social and Political Elements of Climate Change*, 113–32. Berlin: Springer, 2011. doi:10.1007/978-3-642-14776-0_8.

Saad, Lydia, and Jeffrey M. Jones. "U.S. Concern about Global Warming at an Eight-Year High." Gallup.com, March 16, 2016. http://news.gallup.com/poll/190010/concern-global-warming-eight-year-high.aspx.

Sampson, Robert J., Stephen W. Raudenbush, and Felton Earls. "Neighborhoods and Violent Crime: A Multilevel Study of Collective Efficacy." *Science* 277, no. 5328 (August 15, 1997): 918–24. doi:10.1126/science.277.5328.918.

Sartwell, Crispin. "Knowledge Is Merely True Belief." *American Philosophical Quarterly* 28, no. 2 (1991): 157–65.

Scannell, Leila, and Robert Gifford. "Personally Relevant Climate Change: The Role of Place Attachment and Local versus Global Message Framing in Engagement." *Environment and Behavior* 45, no. 1 (2013): 60–85.

Schellnhuber, Hans Joachim, William Hare, Olivia Serdeczny, Sophie Adams, Dim Coumou, Katja Frieler, Maria Martin, et al. *Turn Down the Heat: Why a 4°C Warmer World Must Be Avoided*. Washington, DC: World Bank, 2012. https://www.pik-potsdam.de/.../turn-down-the-heat-executive-summary-english.pdf.

Schuldt, Jonathon P., Sara H. Konrath, and Norbert Schwarz. "'Global Warming' or 'Climate Change'? Whether the Planet Is Warming Depends on Question Wording." *Public Opinion Quarterly* 75, no. 1 (2011): 115–24.

Schuldt, Jonathon P., Katherine A. McComas, and Sahara E. Byrne. "Communicating about Ocean Health: Theoretical and Practical Considerations." *Philosophical Transactions of the Royal Society B* 371, no. 1689 (2016): 20150214.

Schuldt, Jonathon P., Adam R. Pearson, Rainer Romero-Canyas, and Dylan Larson-Konar. "Brief Exposure to Pope Francis Heightens Moral Beliefs about Climate Change." *Climatic Change* 141, no. 2 (March 1, 2017): 167–77. doi:10.1007/s10584-016-1893-9.

Schuldt, Jonathon P., and Sungjong Roh. "Media Frames and Cognitive Accessibility: What Do 'Global Warming' and 'Climate Change' Evoke in Partisan Minds?" *Environmental Communication* 8, no. 4 (October 2, 2014): 529–48. doi:10.1080/17524032.2014.909510.

Schultz, P. Wesley, and Florian G. Kaiser. "Promoting Pro-Environmental Behavior." The Oxford Handbook of Environmental and Conservation Psychology, September 28, 2012. doi:10.1093/oxfordhb/9780199733026.013.0029.

Schusler, Tania M., and Marianne E. Krasny. "Environmental Action as Context for Youth Development." *Journal of Environmental Education* 41, no. 4 (January 1, 2010): 208–23.

Schwartz, Shalom. "Are There Universal Aspects in the Structure and Contents of Human Values?" *Journal of Social Issues* 50, no. 4 (1994): 19–45.

Simmons, Bora, Ed McCrea, Andrea Shotkin, Drew Burnett, Kathy McGlauflin, Richard Osorio, Celeste Prussia, Andy Spencer, and Brenda Weiser. *Nonformal Environmental Education Programs: Guidelines for Excellence*. North American Association for Environmental Education, 2009. https://naaee.org/sites/default/files/nonformalgl.pdf.

Smith, Eliot R., and Diane M. Mackie. "Attitudes and Attitude Change." In *Social Psychology*, 3rd ed., 228–67. New York: Psychology Press, 2007.

Snyder, C. R, Cheri Harris, John R. Anderson, Sharon A. Holleran, Lori M. Irving, Sandra T. Sigmon, Lauren Yoshinobu, June Gibb, Charyle Langelle, and Pat Harney. "The Will and the Ways: Development and Validation of an Individual-Differences Measure of Hope." *Journal of Personality and Social Psychology* 60, no. 4 (1991): 570–85.

Sorenson, Raymond P. "Eunice Foote's Pioneering Research on CO_2 and Climate Warming." *Search and Discovery* article 70092 (2011).

Spence, Alexa, and Nick Pidgeon. "Framing and Communicating Climate Change: The Effects of Distance and Outcome Frame Manipulations." *Global Environmental Change* 20, no. 4 (2010): 656–67.

Spence, Alexa, Wouter Poortinga, and Nick Pidgeon. "The Psychological Distance of Climate Change." *Risk Analysis* 32, no. 6 (June 1, 2012): 957–72. doi:10.1111/j.1539-6924.2011.01695.x.

Stapleton, Sarah Riggs. "Environmental Identity Development through Social Interactions, Action, and Recognition." *Journal of Environmental Education* 46, no. 2 (2015): 94–113.

Stehfest, Elke, Lex Bouwman, Detlef P. van Vuuren, Michel G. den Elzen, Bas Eickhout, and Pavel Kabat. "Climate Benefits of Changing Diet." *Climatic Change* 95, nos. 1–2 (July 2009), 83–102.

Stern, Marc J., and Kimberly J. Coleman. "The Multidimensionality of Trust: Applications in Collaborative Natural Resource Management." *Society and Natural Resources* 28 (2015): 117–32.

Stern, Paul C., Thomas Dietz, Troy Abel, Gregory A. Guagnano, and Linda Kalof. "A Value-Belief-Norm Theory of Support for Social Movements: The Case of Environmentalism." *Research in Human Ecology* 6, no. 2 (1999): 81–97.

Stevenson, Kathryn T., Marcus A. Lashley, M. Colter Chitwood, M. Nils Peterson, and Christopher E. Moorman. "How Emotion Trumps Logic in Climate Change Risk Perception: Exploring the Affective Heuristic among Wildlife Science Students." *Human Dimensions of Wildlife* 20, no. 6 (2015): 501–13.

Stevenson, Kathryn T., and Nils Peterson. "Motivating Action through Fostering Climate Change Hope and Concern and Avoiding Despair among Adolescents." *Sustainability* 8, no. 1 (January 2016): 6. doi:10.3390/su8010006.

Stevenson, Kathryn T., M. Nils Peterson, and Howard D. Bondell. "The Influence of Personal Beliefs, Friends, and Family in Building Climate Change Concern among Adolescents." *Environmental Education Research*, 2016, 1–14.

Swim, Janet K., and John Fraser. "Fostering Hope in Climate Change Educators." *Journal of Museum Education* 38, no. 3 (2013): 286–97.

Swim, Janet K., Nathaniel Geiger, John Fraser, and Nette Pletcher. "Climate Change Education at Nature-Based Museums." *Curator: The Museum Journal* 60, no. 1 (January 1, 2017): 101–19. doi:10.1111/cura.12187.

Taylor, Dorceta E. *The State of Diversity in Environmental Organizations: Mainstream NGOs, Foundations, Government Agencies*. Green 2.0, 2014. http://vaipl.org/wp-content/uploads/2014/10/ExecutiveSummary-Diverse-Green.pdf.

Tyndall, John. "On the Absorption and Radiation of Heat by Gases and Vapours, and on the Physical Connexion of Radiation, Absorption and Conduction." *Philosophical Transactions of the Royal Society of London* 151, no. 1 (1861): 1–36.

Uhl, Isabella, Johannes Klackl, Nina Hansen, and Eva Jonas. "Undesirable Effects of Threatening Climate Change Information: A Cross-Cultural Study." *Group Processes and Intergroup Relations*, 2017, 1–17.

United Nations Framework Convention on Climate Change (UNFCCC). "Gas Emissions from Waste Disposal." 2002. http://www.grid.unep.ch/waste/download/waste4243.PDF.

University Corporation for Atmospheric Research (UCAR). "Climate Change and Vector-Borne Disease." 2011. https://scied.ucar.edu/longcontent/climate-change-and-vector-borne-disease.

——. "How Volcanoes Influence Climate." 2017. https://scied.ucar.edu/shortcontent/how-volcanoes-influence-climate.

U.S. Global Change Research Program (USGCRP). *Climate Science Special Report: A Sustained Assessment Activity of the U.S. Global Change Research Program.* 2017.

Volmert, Andrew. "Getting to the Heart of the Matter: Using Metaphorical and Causal Explanation to Increase Public Understanding of Climate and Ocean Change." FrameWorks Institute, May 2014. http://www.frameworksinstitute.org/assets/files/PDF_oceansclimate/occ_metaphor_report.pdf.

Waldman, Scott. "Maryland Island Denies Sea Level Rise, Yet Wants to Stop It." *Scientific American*, ClimateWire, June 15, 2017. https://www.scientificamerican.com/article/maryland-island-denies-sea-level-rise-yet-wants-to-stop-it/.

Walker, Brian, and David Salt. *Resilience Thinking: Sustaining Ecosystems and People in a Changing World.* Washington, D.C.: Island Press, 2012.Washburn, Anthony N., and Linda J. Skitka. "Science Denial across the Political Divide: Liberals and Conservatives Are Similarly Motivated to Deny Attitude-Inconsistent Science." *Social Psychological and Personality Science*, September 2017. doi:10.1177/1948550617731500.

Watts, Nick, Markus Amann, Sonja Ayeb-Karlsson, Kristine Belesova, Timothy Bouley, Maxwell Boykoff, Peter Byass, et al. "The *Lancet* Countdown on Health and Climate Change: From 25 Years of Inaction to a Global Transformation for Public Health. *Lancet.* doi:10.1016/S0140–6736(17)32464–9.

Wells, David A. "Heat of the Sun's Rays" (1857). In *Annual of Scientific Discovery: Or, Year-Book of Facts in Science and Art for 1857*, edited by David Wells, 159–160. Boston: Gould and Lincoln.

Whitmarsh, Lorraine. "What's in a Name? Commonalities and Differences in Public Understanding of 'Climate Change' and 'Global Warming.'" *Public Understanding of Science*, 2008. http://pus.sagepub.com/content/early/2008/09/16/0963662506073088.short.

Wibeck, Victoria. "Enhancing Learning, Communication and Public Engagement about Climate Change—Some Lessons from Recent Literature." *Environmental Education Research* 20, no. 3 (2014): 387–411.

Williams, Corrie Colvin, and Louise Chawla. "Environmental Identity Formation in Nonformal Environmental Education Programs." *Environmental Education Research* 22, no. 7 (2015): 978–1001.

World Meteorological Organization. "Greenhouse Gas Concentrations Surge to New Record." October 30, 2017. https://public.wmo.int/en/media/press-release/greenhouse-gas-concentrations-surge-new-record.

Worth, Katie. "Climate Change Skeptic Group Seeks to Influence 200,000 Teachers." *Frontline*, March 28, 2017. https://www.pbs.org/wgbh/frontline/article/climate-change-skeptic-group-seeks-to-influence-200000-teachers/.

Yale Program on Climate Change Communication. "What Is Climate Change Communication?" Accessed July 9, 2017. http://climatecommunication.yale.edu/about/what-is-climate-change-communication/.

——. "Yale Climate Opinion Maps—U.S. 2016." Accessed January 3, 2018. http://climatecommunication.yale.edu/visualizations-data/ycom-us-2016/.

Index

Note: Page numbers in *italics* indicate figures; those with a *t* indicate tables.

CPSIA information can be obtained
at www.ICGtesting.com
Printed in the USA
LVHW04s0113300918
591802LV00002BA/102/P